The Assyrian Church of the East

The Assyrian Church of the East

History and Geography

Christine Chaillot

Peter Lang
Oxford · Bern · Berlin · Bruxelles · New York · Wien

Bibliographic information published by Die Deutsche Nationalbibliothek
Die Deutsche Nationalbibliothek lists this publication in the Deutsche
Nationalbibliografie; detailed bibliographic data is available on the Internet at
http://dnb.d-nb.de.

A catalogue record for this book is available from the British Library.

Library of Congress Cataloging-in-Publication Data

Names: Chaillot, Christine, author.
Title: The Assyrian Church of the East : history and geography / Christine Chaillot.
Description: Oxford ; New York : Peter Lang, [2021] | Includes bibliographical references and index.
Identifiers: LCCN 2020054265 (print) | LCCN 2020054266 (ebook) | ISBN 9781789979121 (paperback) | ISBN 9781789979138 (ebook) | ISBN 9781789979145 (epub) | ISBN 9781789979152 (mobi)
Subjects: LCSH: Assyrian Church of the East--History. | Church of the East--History. | Asia--Church history.
Classification: LCC BX153.3 .C43 2021 (print) | LCC BX153.3 (ebook) | DDC 281/.8--dc23
LC record available at https://lccn.loc.gov/2020054265
LC ebook record available at https://lccn.loc.gov/2020054266

Translation by Norman Russell
Cover illustration: the patriarchal church of St Shallita at Kochanes (south-east Turkey) in the nineteenth century.
Cover design by Brian Melville for Peter Lang.

ISBN 978-1-78997-912-1 (print) • eISBN 978-1-78997-913-8 (ePDF)
ISBN 978-1-78997-914-5 (ePub) • ISBN 978-1-78997-915-2 (mobi)

© Peter Lang Group AG 2021

Published by Peter Lang Ltd, International Academic Publishers,
52 St Giles, Oxford, OX1 3LU, United Kingdom
oxford@peterlang.com, www.peterlang.com

Christine Chaillot has asserted her right under the Copyright, Designs and Patents Act, 1988, to be identified as Author of this Work.

All rights reserved.
All parts of this publication are protected by copyright.
Any utilisation outside the strict limits of the copyright law, without the permission of the publisher, is forbidden and liable to prosecution.

This applies in particular to reproductions, translations, microfilming, and storage and processing in electronic retrieval systems.

This publication has been peer reviewed.

By the same author:

The Malankara Orthodox Church, Geneva, 1996

Towards Unity, The Theological Dialogue Between the Orthodox Church and the Oriental Orthodox Churches, Geneva, 1998

The Syrian Orthodox Church of Antioch, Geneva, 1998

A Short History of the Orthodox Church in Western Europe in the Twentieth Century, Paris, 2006

The Orthodox Church in Eastern Europe in the Twentieth Century, Oxford, Peter Lang, 2011

Vie et spiritualité des Églises orthodoxes orientales des traditions syriaque, arménienne, copte et éthiopienne, Paris, Les Éditions du Cerf, 2010

Les Coptes d'Égypte. Discriminations et persécutions (1970–2011), Paris, L'Harmattan, 2014

The Dialogue Between the Eastern Orthodox and Oriental Orthodox Churches, Volos Academy Publications, Volos, 2016

The Role of Images and the Veneration of Icons in the Oriental Orthodox Churches, LIT Verlag, Münster, 2017

'Many monks [from the Church of the East] have crossed the seas and have gone to India and China, carrying only their staff and their pouches as luggage.'
Letters of Patriarch Timothy I (780–823), in R. Badawi (ed.), Studi e Testi 187, Vatican, 1956.

'By the 7th century Persian missionaries had reached the "end of the world", the capital of Tang-dynasty China.'
S. Moffett, A History of Christianity in Asia, Beginnings to 1500, San Francisco, 1992, p. xv

'Few churches can claim for themselves the Nestorian evangelizing fire that swept all over the continent of Asia in the earlier Middle Ages ... It is no exaggeration to say that in the early Middle Ages the Nestorian Church was the most widespread in the whole world.'
A. S. Atiya, A History of Eastern Christianity, London, 1968, pp. 240 and 287.

'The suffering of the Assyrian and other Syriac Christians in the Ottoman Empire during World War I is one of the least known genocide of modern times, a genocide that targeted also Armenians.'
D. Gaunt, 'Failed Identity and the Assyrian Genocide', in Shatterzone of Empires, 2013, pp. 317–33.

'On 20 November 1937, Assyrian Patriarch Eshai Shimun XXI, then in London, sent a long message to his people : "Twenty-two years have passed since the loss of the land of our forefathers [the genocide of 1915] and four years since the bitter experience in Iraq [Simele, in 1933]. During this time, all our possessions and no less than the three-fifths of our people were lost ... With all the faculties of our body and soul, we must throw ourselves into the struggle to save our nation from annihilation, struggling by the most peaceful means, animated by hope".
C. Weibel Yacoub, Le rêve brisé des Assyro-Chaldéens. L'introuvable autonomie, Paris, 2011, pp. 279–80.

'My people have had a great struggle to maintain their Christian faith ... But their witness goes on, and I pray God that easier days may soon be granted them.'
Patriarch Eshai Shimun XXI of Church of the East, Foreword to his book The Assyrian Tragedy, 1934.

'... Christians have systematically forgotten or ignored so very much of their history that is scarcely surprising that they encounter only a deafening silence. Losing the ancient churches is one thing, but losing their memory and experience so utterly is a disaster scarcely less damaging. To break the silence, we need to recover those memories, to restore that history.'
Philip Jenkins, The Lost History of Christianity, Oxford, 2008, p. 262.

Contents

Preface by Dr Sebastian Brock   xi

Introduction   1

CHAPTER 1
A History of the Church of the East: Origins to the Eighteenth Century   9

CHAPTER 2
In Arabia and the Persian Gulf   35

CHAPTER 3
In India   43

CHAPTER 4
In Central Asia and Beyond. On the Silk Road   55

CHAPTER 5
In China under the Tang (635–845) and the Mongols (1206–1368)   71

CHAPTER 6
Under the Mongols (1206–1368) and Tamerlane (1370–1405)   85

CHAPTER 7
The Nineteenth Century   101

CHAPTER 8
The Twentieth Century   111

CHAPTER 9
The Twenty-First Century around the World. The Diaspora   129

Conclusion   145

Bibliography   151

Timeline   187

Maps   193

Preface

When Eusebius of Caesarea wrote the first History of the Christian Church at the beginning of the fourth century he unwittingly set a precedent for writing about the History of the Church which has had some very unfortunate consequences. What Eusebius wrote was in fact a history of the Church *within the confines of the Roman Empire*, leaving out its history in the adjacent Persian Empire, further to the east. This was perfectly understandable from Eusebius' own perspective: what is most unfortunate, however, is that this limitation of the history of the early Church to the confines of the Roman Empire has all too often been followed, right up to the present day. As a consequence the fascinating history of the Church that sprung up to the east of the Roman Empire, and which is represented today by the Church of the East, is very little known to the wider public. In view of this situation, Christine Chaillot's new book is most timely and greatly to be welcomed.

The tradition of the Church of the East is in fact represented today by three separate Churches: the Assyrian Church of the East, the Ancient Church of the East and the Chaldean Catholic Church. While the roots of the last of these go back to the mid-sixteenth century, the break between the first two only came in the last half of the twentieth century, primarily over the issue of the calendar, with the Assyrian Church of the East adopting the New (Gregorian) Calendar and the Ancient Church of the East retaining the Old (Julian) Calendar. Christine Chaillot's prime focus of attention is on the Assyrian Church of the East and the Ancient Church of the East.

The negative effects of Eusebius' limited model for the subsequent writing of Church History have ensured that the remarkable expansion of the Church of the East until the fourteenth century is hardly known outside specialist circles. Dramatic witnesses to this expansion are the Xian Stele with its Chinese-Syriac inscription dated AD 781, and the large collection of manuscript fragments of texts of the Church of the East (ninth to fourteenth century) in different languages, discovered in the early twentieth century in Turfan (north-western China).

Knowledge about, and a proper appreciation of, the role of the Church of the East within the wider history of the Christian Church, has suffered from a second disadvantage. The Christological controversies of the fifth and sixth centuries had produced a three-way split in Christianity which continues to this day: (1) the position of the Greek Byzantine East, Latin West and later Reformed traditions; (2) that of the Oriental Orthodox Churches; and (3) that of the Church of the East. Only the first position accepts the Doctrinal Formula of the Council of Chalcedon (451). While both the first and the third can be described as 'dyophysite', since they speak of two 'natures' (*physeis*) as existing in the incarnate Christ, the Oriental Orthodox hold that there is 'a single <u>composed</u> nature' (*mia physis*), and so may conveniently be designated as 'miaphysite'. In the heat of these theological controversies, each side tried to associate their opponents with names considered heretical, and over the course of time many of these sobriquets came to stick. It has only been in recent times that serious attempts have been made to get rid of misleading epithets, one of which, 'Nestorian' was very widely used of the Church of the East, thus associating the Church closely with Nestorius, patriarch of Constantinople, who was the archopponent of Cyril of Alexandria, and who was deposed by the Council of Ephesus in 431. From the present standpoint, it is essential to realize that the name 'Nestorius' means three very different things: (1) for the Oriental Orthodox and Eastern Orthodox the name implies a seriously heretical position that disassociates the Son of the Virgin Mary from the Son of God and so divides up Christ; (2) to the Church of the East 'Nestorius' is an upholder of the dyophysite Christological position against Cyril of Alexandria, and at the same time the putative author of one of the three Anaphoras in use in the Liturgy of the Church; (3) for modern scholars the real theological position of Nestorius is a matter of continuing dispute, in view of the tendentious nature of almost all the surviving sources concerning him and the consequent difficulty in interpreting them globally.[1]

---

1   Cf. my 'The "Nestorian" Church: lamentable misnomer', in the *Bulletin of the John Rylands Library, Manchester* 78 (1996), pp. 23–36. Needless to say, the Church of

Preface by Dr Sebastian Brock

Since the dominant narrative of Church History has been that of the Chalcedonian Churches' authors, the Church of the East has regularly, but utterly misleadingly, been designated as 'Nestorian'. The nomenclature continued into the time of Islamic domination, and it is only in recent times that it has rightly been challenged, both in academic writing and in ecumenical discussion. Especially significant in drawing attention to past misconceptions of the Christology of the Church of the East has been the Common Statement on Christology between Pope John Paul II and Patriarch Mar Dinkha IV of the Church of the East in November 1994, and the more recent 'Common Statement on the Sacramental Life' between the Vatican and the Assyrian Church of the East, in November 2017.[2] These initiatives should be considered as an important starting point for future dialogue between the Assyrian Church of the East with other Churches of the Chalcedonian tradition, and in particular the Eastern Orthodox and Oriental Orthodox Churches. The proper spirit for the undertaking of such a dialogue was nicely indicated by the great Syriac Orthodox polymath, Barhebraeus, already in the late thirteenth century:

> When I had given much thought and pondered on the matter, I became convinced that these quarrels of Christians are not a matter of factual substance, but rather, one of words and terms; for they all confess Christ our Lord to be perfect God and perfect human, without any commingling, mixing or confusion of the natures. This bi-pinnate likeness[3] is termed by one party [i.e. the Oriental Orthodox Churches] a 'nature', by another [i.e. the Chalcedonian Churches] a 'hypostasis (person)', and by yet another [i.e. the Church of the East] a 'prosopon (*parsopa*, person)'. Thus I saw all the Christian communities, with their different Christological positions, as possessing a single common ground that is without any difference. Accordingly, I totally eradicated any hatred from the depths of my heart and I completely renounced disputing with anyone over confessional matters.
>
> (Barhebraeus, *Book of the Dove*, chapter 4)

---

    the East does not in fact hold the heretical positions often attributed to them in the polemical literature.

2   The main section of the 'Common Declaration of Faith' of 1994 can be found in my 'The Syriac Churches in Ecumenical Dialogue on Christology', in A. O'Mahony (ed.), *Eastern Christianity. Studies in Modern History, Religion and Politics* (London, 2004), pp. 44–65, here pp. 54–5.

3   The Syriac is *dmutha* for likeness ; and the second word, *suniptronayta*, is a borrowing from Greek *sunipteros* ('flying with both wings').

It can be seen that, for a variety of most unfortunate historical reasons, the remarkable history of the Church of the East has been largely overlooked or, where this has not been the case, it has usually been misrepresented and misunderstood. Accordingly, all the greater is the need for a reliable, and at the same time sympathetic, short history of the Church of the East written for the benefit of a wider reading public: this is precisely what Christine Chaillot has aimed to provide, and it is greatly to be hoped that her attractive book will help to dispel both the general ignorance about, and the all too many misconceptions concerning, this venerable ancient Church.

Sebastian Brock

The Oriental Institute, Oxford University

Introduction

This book is an introduction to the history and geography of the Church of the East, the two subjects being incapable of separation. On the one hand, the history cannot be understood without the places; on the other, without the history, it is impossible to understand the situation as it exists today.

The missionary journeys of the apostle Paul at the beginning of Christian history, starting from Antioch (today the town of Antakia in south-east Turkey), continuing into Greece, and ending up in Rome (Acts 13: 1–4), are well known. But the history of Christianity in the East, not only in the Middle East but beyond it in Asia, is often ignored. It is also from Antioch that Christianity spread eastwards to Edessa (the modern town of Urfa in south-east Turkey, 350 km north-east of Antioch) and in consequence to the whole of the Persian Empire. It was in Mesopotamia (between the Tigris and the Euphrates) that the first centre of the Church of the East developed, at Seleucia-Ctesiphon, the capital of the great Persian Empire (today to the south of Baghdad), and it was from there that this Church played a very important role in the spread of Christianity towards Eastern Asia.

The name 'the Church of the East' suits it very well, as it is situated to the east of the other Middle-Eastern Churches. From the beginning of Christianity until the fourteenth century, this Church experienced an extraordinary missionary drive and a remarkable expansion in Asia. It brought the Gospel to the whole of the East and even to the Far East, from Persia as far as India, via the Persian Gulf, and even as far as China, which it reached in the seventh century via Central Asia and the celebrated Silk Road that linked China to the Mediterranean world. In this book I have arranged the order of chapters in accordance with these itineraries. There were other Churches in Asia during these centuries, but we shall concentrate here on the Church of the East.

It was a combination of religious and political factors, the Mongol conquest (1206–1368), the reign of the Yuan dynasty in China (which ruled until 1368), and above all the era of the patriarch Yahballaha III (1281–1317), that permitted the Church of the East to live through one of the most glorious periods of its history, a period comparable to the one which it had experienced in the past under the patriarch Timothy I (780–823). It was at that time that the Church of the East reached its maximum territorial expansion.

In Persia, during Mongol rule (the *ilkhanate*), Islam became the official religion under Ghazan (1295–1304). Then the invasions of Timur (Tamerlane) (1370–1405) right across Asia brought about the great decline of the Church of the East. As a result of his devastating conquests, most of the dioceses of the Church of the East were destroyed. In consequence, only the communities which found refuge in remote regions in the north of Mesopotamia were able to survive. These regions were in the mountains difficult of access lying between the north of Iraq and Lake Van, the Hakkari mountains (in the south-east of modern Turkey), where the patriarch resided at Qochanis until 1915. There still existed at that time small communities in neighbouring Persia (around Urmia, in the north-west of modern Iran), and in south-west India, at Kerala. After the genocide suffered by Christians in the Ottoman empire in 1915, and the subsequent massacres in Persia, the Church of the East and its people were on the verge of extinction. In consequence of these events, the Assyrian patriarch Shimun XXI Eshai (1920–75) saw the death of three-fifths of his faithful, between 60,000 and 72,000 people. Since 2015 the seat of the Church of the East has been at Erbil, in the north of Iraq. Many of the faithful have left the Middle East and have formed diaspora communities throughout the world.

Within the perspective of the general history of Christianity, the Church of the East is one of the earliest Churches. Subject at first to the jurisdiction of the Church of Antioch, it proclaimed its independence in 424 at a local synod. It must be emphasized that until the fifth century there were no divisions within the Church. It was in the aftermath of the Council of Ephesus in 431 that the first schism came about, with the separation of the Church of the East. Subsequently, there was a second division after the Council of Chalcedon of 451 with the Churches which for too long have wrongly been called 'monophysite' and are today referred to as the 'Oriental Orthodox Churches', namely, the Churches of the Coptic, Ethiopian, Armenian and Syrian Orthodox traditions. Then there occurred the great schism between East and West in 1054, when the Orthodox Churches of the Byzantine rite and the Catholic Church of Rome went their separate ways. As for the Protestant Churches, their existence goes back to the sixteenth century.

On the theological level, the Church of the East did not accept the Council of Ephesus. For centuries the Christology of this Church has been associated with that of Nestorius, the patriarch of Constantinople (428–31) condemned as a heretic at the Council of Ephesus, and the Church of the East has been characterized as 'Nestorian'. As a result of extensive scholarly research, modern theologians and historians have concluded that this condemnation lacks justification. This is discussed in the Preface by Dr Sebastian Brock, who writes elsewhere that 'to continue to call that Church "Nestorian" is, from the historical point of view, totally misleading and incorrect – quite apart from being highly offensive and a breach of ecumenical good manners'. The appellation 'Nestorian' is officially rejected by the Church of the East. We must respect this and refrain from using the term.

The Church of the East is a Church with a Syriac literary tradition and liturgical rite. Syriac is a Christian language of Semitic origin. It was the local Aramean dialect of Edessa. Jesus Christ spoke the Aramean of Palestine. At the beginning of Christianity, Syriac was widely spoken in the Middle East, particularly in the region of Edessa. A classical language, like Latin in the West, it was one of the important languages of ancient Christianity and a language of high culture in the Middle East, especially in Syria and Mesopotamia. Even some of the Christians of the patriarchate of Antioch calling themselves 'Greek Orthodox' spoke, wrote and prayed in Syriac. Numerous literary works were written in this language. It was only after the arrival of Islam in the seventh century that Syriac Christians also began to speak and write in Arabic.

The language of the Christians of the Church of the East and of the Chaldean Catholic Church is called 'Eastern' Syriac. The Eastern Syriac dialect is still spoken today and is called Sureth. There is another group of Churches of Syriac tradition and rite: the Syrian Orthodox Church, the Syrian Catholic Church and also the Maronite Church (in Lebanon, united with Rome), which are of the tradition called 'Western Syriac'. Eastern Syriac is a little different from Western Syriac in its writing and in certain forms of pronunciation, using, for example, the sound 'a' in place of 'o' at the end (and sometimes in the middle) of certain words and names. One may regard all these Churches as belonging to a single large family of Churches of Syriac origin, for this linguistic bond unites them strongly on a cultural level.

The names accorded to these different Churches of the Syriac rite can cause confusion; they can even change from one country to another. To understand them correctly we need to study them carefully. Since 1976, the official name 'the Church of the East' has been 'the Holy Apostolic Catholic Assyrian Church of the East' ('catholic' meaning 'universal'). In this book, it will be abbreviated as the Assyrian Church of the East and also as the Church of the East. The head of the Church bears the title 'catholicos-patriarch', abbreviated here to 'patriarch'. Since 1968, there has been a second patriarchate of the same tradition, with its seat at Baghdad, called the Ancient Church of the East, which is identical with the Church of the East from which it derives except for the distinctive feature of following the Old (Julian) Calendar. Just like the Church of the East, it has parishes and members in the Middle East and throughout the world. The Church of the East is the mother-church of the Chaldean Catholic Church, which has been subject to the Catholic Church of Rome since 1553. In India the Church of the East is called the Chaldean Syrian Church of the East, and recently also the Assyrian Church of the East, which is a source of confusion, and the Catholic branch is called the Syro-Malabar Church. To complicate matters further, in India the Syrian Orthodox of the Patriarchate of Antioch have formed two churches: one is called the 'Malankara Jacobite Syrian Orthodox Church', dependent on the Patriarchate of Antioch, whose seat is today at Damascus; the other is autocephalous with its seat at Kottayam in India and is called the 'Malankara Orthodox Syrian Church of India' (also called today the 'Malankara Indian Orthodox Church'); the parallel Catholic branch in India is called the 'Syro-Malankar Church'.

Some call this group of Churches of Syriac tradition and their people 'Assyrians', in order to emphasize their common Syriac roots and their history in Mesopotamia, in the territory of ancient Assyria. It was in this region in the second millennium BC that a powerful kingdom arose which later became an empire. In the eighth and seventh centuries BC Assyria controlled territories extending over all or most of the modern countries of Iraq, Syria, Lebanon, Iran and Turkey. Assyria is the cradle of one of humanity's first great civilizations. In fact, in antiquity, before the birth of Christianity, the Mesopotamian cities of Nineveh and Babylon contributed to the glorious history of the ancient Sumerians, Babylonians, Chaldeans and Assyrians. It may be mentioned, too, that the Assyrian Empire features in the narratives of the Old Testament.

The name 'Assyrian' has been used since the nineteenth century as the name of the Christians who follow the tradition of the Church of the East, which explains why I use this name for them from Chapter 7 onwards. Since the First World War (1914–18) it has also been used in the political domain to designate the nation or ethnic group represented historically by the Church of the East and the other Churches of Syriac tradition listed above. In this book I shall use the term 'Assyro-Chaldeans' to designate at the same time both the Assyrians of the Church of the East and the Chaldean Catholics.

At the present time, the name 'Assyrians' has acquired a broader reference in certain Christian universities, and political and media circles, for example, the Assyrian International News Agency (AINA), so as to include all the Christian communities of the Syriac tradition. This allows problems to be emphasized and connected, among others, with the history and geopolitics common to the Syriac Christian communities. But in my view this can equally lead to confusion if one wishes to refer to one of these groups or one of these Churches in particular.

As Arthur Goldschmidt writes: 'History depends upon geography. Before you can have a play, there must be a stage. Let us stick to the few essential points that one needs to master before starting a study.' To explore the history of the Church of the East in a serious way, you have to set it within its geographical context, both past and present. For this reason, so as to make the book easier to read, I include maps which locate a very large number of place-names mentioned in this book. These maps are based on those given by David Wilmshurst (2011), Françoise Briquel Chatonnet and

Muriel Debié (2017), and Daniel King (2019). Google Maps may also be consulted. For the spelling of geographical names, I have tried to follow the forms generally used in English; for certain Chinese names, I have followed the spelling used by David Wilmshurst in his book *The Martyred Church* (the Wade Giles system) together with the Pinyin version (the transcription of Chinese using the Latin alphabet, which is the most widespread version used today in modern works), thanks to the help of Li Tang.

To the historical and geographical information I should like to add some details on the principal archaeological sites of the countries which concern us, notably in Central Asia and modern China and also in the region of the Persian Gulf. Archaeology witnesses to the presence of the Church of the East in Asia: the excavations that have made possible the rediscovery of certain monasteries and churches, and, moreover, inscriptions incised on steles and tombstones are tangible and irrefutable proof of this Church's history, including places where one does not expect to find it. Christoph Baumer's book, *The Church of the East* (2016), gives numerous examples with photographs.

Because my work is presented as a general overview, there are no footnotes. To compensate for this, I have prepared a bibliography of the numerous works I have consulted, presented chapter by chapter, so that the reader may go more deeply into the subject. One should also consult the articles and bibliographies in the very recent book by Daniel King, *The Syriac World* (2019), which includes links to numerous online resources. Some of the maps, as well as some of the books and articles cited in the Bibliography, are available on the internet (e.g. at <https://www.academia.edu/>) occasionally with photographs.

One of the difficulties I encountered in my work concerned the choice of dates, on which there is a lack of scholarly unanimity, for there are numerous research projects still in progress. Out of concern for consistency, I decided as a rule to follow the dates given by the historian David Wilmshurst in his book *The Martyred Church: A History of the Church of the East* (2011), using, for example, his list of the patriarchs of the Church of the East. This book, whose title calls to mind the persecutions suffered by the Church of the East throughout the centuries, is my principal reference work from several angles, particularly for the many historical references as well as for a number of statistics – another controversial subject.

Furthermore, this book has been conceived in such a way that each chapter can be read independently of the others. The reader will notice a gap between the fifteenth and the nineteenth centuries. This is a period of great decline for the Church of the East, the details of which are still poorly understood and are the subject of current research by specialists. I have therefore omitted this period from my narrative. A general timeline will be found at the end of the book. Certain proper names are given in their Syriac form or in their English equivalent, for example, Shimun/Simon, Eshai/Isaiah and Giwargis/George. Dinkha is the pronunciation in the East Syriac dialect of the name 'Denha'. The word 'Mar', which is the Syriac word for 'Lord', may precede the name of a saint, a patriarch or a bishop.

According to the Lebanese historian, Georges Corm, 'history is the study of the complexity of events; to give a historical account is always very complex and demands a great deal of painstaking research; to write history with objectivity and respect is a challenge, for our history is not that of others and vice versa'. We should never forget these wise words which help us understand the limitations of every historian and writer.

It only remains for me to invite the reader to travel in time and space, to undertake the fascinating discovery of the Church of the East, this very ancient apostolic Church whose history constitutes an indispensable chapter of the history of the universal Church.

In conclusion, I should like to thank all who have helped me write this book by their advice and support. I cannot list all their names here, but among them I must mention, in particular, several bishops of the Church of the East and the patriarch Giwargis III. I am most grateful, too, to Dr Sebastian Brock for his Preface, and to Dr Norman Russell for his translation.

CHAPTER 1

A History of the Church of the East: Origins to the Eighteenth Century

The First Centuries

Christianity appeared in the Roman Empire, which incorporated both the eastern and western parts of the Mediterranean world, with its capital, Rome, laying in the western part. There is a tendency to forget the fact that in the beginning the principal centres of Christianity were situated not only in Jerusalem and Antioch, but also in Edessa, an ancient city of Upper Mesopotamia.

It was at Antioch (today in south-east Turkey) that Christ's disciples first received the name of Christians (Acts 11:19–26). Antioch, an apostolic see whose first primate was St Peter before he went to Rome, was a very important centre for the spread of Christianity in the region. Originally the Church of the East was tied to the see of Antioch, but its presence in Persian, not Byzantine, territory led, among other things, to its autonomy.

The city of Edessa (today Urfa in Turkey, about 260 km north-east of Antioch) and its environs were places principally of Syriac language and culture and remained so. Edessa at that time was situated on the crossroads of the commercial routes between the Roman and Parthian Empires. From Edessa Christianity spread eastwards into the Parthian Empire in Persia (an area covering modern Iran, Iraq and surrounding lands). It was in the Persia of the Parthians (247 BC–224 AD) and then of the Sasanids (224–637) that the Church of the East developed independently of the Churches situated in the West.

In the Persian Empire, Christianity grew as a result of conversions among the Persian population and the conversion of a number of Jews, but equally through the arrival in Persia of Christian merchants, along with a number of Christian refugees fleeing Roman persecution, or even Christians deported as prisoners in the wake of Persian conquests.

With regard to the beginning of the general history of Christianity, the day of Pentecost is considered as marking the birth of the Church. Pentecost took place in Jerusalem, then in the territory of the Roman Empire. It is claimed that the people present at this event came from neighbouring countries, including 'Parthians, Medes, Elamites, and residents of Mesopotamia' (Acts 2:9). The Parthian Empire was then a major political power and the main rival of the Roman Empire. The Medes were an ancient Persian people who inhabited a region called Medea (today in north-west Iran). The Elamites had their historical centre in the south-west of modern Iran, to the east of the lower Tigris. Elam is called Beth Huzaye in Syriac and is today situated in the Iranian province of Khuzistan, in the area round the city of Gundishapur (in Syriac Beth Lapat), an area that was subsequently to be the territory of an important metropolitanate of the Church of the East.

Mesopotamia itself, literally 'the region between two rivers', that is, the Tigris and the Euphrates, lies today principally in modern Iraq. It is in all these places that the Church of the East has its roots. In the Sasanian Persian Empire most of the Christians lived on the western frontier of the empire, in Mesopotamia. In this region, according to Syriac texts and to tradition, the first evangelization is attributed to three apostles, or disciples, of Christ – Addai, Mari and Thomas – who were active in different geographical zones.

Tradition has it that Addai (the Syriac form of the Greek name 'Thaddeus') evangelized the kingdom of Osrhoene, whose capital was Edessa, and converted its king, Abgar V. As the latter suffered from some illness and had heard of healings preformed by Jesus Christ, he sent him a letter asking him to come and heal him. The account of this correspondence (legendary, according to historians) may be read in Eusebius of Caesarea's *Ecclesiastical History* (fourth century) A similar account is found in Syriac in the *Doctrine of Addai* (*c.* 420). Instead of the sending of a letter, Abgar's emissary brought back a portrait of Christ impressed on a textile, which is called the *Mandylion* and is considered a kind of icon. On leaving Edessa,

Addai preached in the environs of the city, in the Parthian Empire, at Beth Zabdai, and in Adiabene (whose ancient capital was Arbela, today Erbil, in northern Iraq).

With regard to Mari, one of Christ's seventy disciples (Luke 10:1–24), he was himself converted by Addai, who sent him to evangelize Mesopotamia and its surrounding area, above all in the south-east, in Babylonian territory. Some consider him the first missionary of the Parthian Empire and the founder of the ecclesiastical seat of the Church of the East at Seleucia-Ctesiphon, the capital of the Parthian Empire (situated about 35 km south-east of Baghdad, on the Tigris).

In the Eastern part of the Persian Empire, the apostle Thomas evangelized the region of Fars, the original land of the Persians (on the Persian Gulf, today in Iran, centred on the city of Shiraz). It is probably from there that he went by sea as far as India.

The book of *The Acts of Mar Mari*, known only in the Syriac language and written after the Sasanian period (between seventh and eighth centuries) in the region of Beth Aramaye (literally 'the land of the Arameans', the ancient Babylonia) sets out in thirty-four tracts the history of the Christianization of the Tigris valley as far as Fars in the east. Starting from Edessa and Nisibis, Mari travelled by way of Arzanene (a historical region in the south-west of Great Armenia to the west of Lake Van, today in eastern Turkey), to Adiabene in the region of the two rivers Zab (the Great and the Lesser), Beth Garmai (a region to the south-west of Kirkuk), and Beth Aramaye, where he preached in the Persian capital of Seleucia-Ctesiphon. His mission ended in Mesene at the mouth of the Tigris and the Euphrates, at the head of the Persian Gulf (near the modern city of Basra, in Iraq), and at Beth Huzaye or Elam, that is to say, at the frontiers of the lands of the Apostle Thomas's mission. *The Acts of Mar Mari* offer a rich documentation relating to Christianity in this region in this period. The text also justified the Church of the East's claim to recognition of its sovereignty and local independence in the Persian Empire. The relics of St Mari were kept in the monastery of Dorqonie, near Seleucia-Ctesiphon.

What can one say about the history of the Church of the East apart from the fact that the sources are complex and conflicting? For historians even the first source (just quoted), *The Acts of Mar Mari*, is regarded as fictitious, even if it provides us with interesting information. This explains why the tables of the early chronology of the Church of the East differ, as it is difficult to define one single clear lineage for the first bishops, and that is why to this day no clear established succession can be given by historians. In principle, this succession comprises the apostle Thomas and three of the seventy disciples – Adai, Aggai and Mari – who were the first to bring the Gospel to Mesopotamia.

In their chronological lists, the eastern chroniclers generally put first the names of Thomas, Addai, Aggai and Mari. In his book, *The Patriarchs of the Church of the East*, Daniel D. Benjamin reveals a great number of variations among the sources and also inconsistencies in the dates of succession of certain patriarchs. As the historians Christelle and Florence Jullien explain, the list of the succession does not always begin in the same way or with the same names. Somes lists give prominence to Addai as being the first occupant of the Church of the East's see at Seleucia-Ctesiphon. Others favour Mari at the expense of Addai. Aggai, for his part, would have been Addai's successor. Others prefer Thomas as the first apostle of the East.

According to Chip Coakley, there are two modern official lists of the Church of the East, the first dating from 1924, the second from 1965, with a reprint in 1988, at the end of which the name of the patriarch Dinkha IV (1976–2015) is added. Both lists begin with Simon Peter (cf. 1 Pet 1:1 and 5:13), followed by Thomas the Apostle and then (but not in the list of 1924) by the apostle Bar Tulmay (Bartholomew). Only then is Addai named, followed by Aggai and then Mari. These successions include the names given in the *Acts of Mar Mari*. With regard to the chronological order of the names, the problem lies in the fact that there are no historical sources that are absolutely reliable for this first period; moreover, by giving these names each local Church wanted to demonstrate its apostolic origins.

Around 280, bishop Papa Bar Aggai was ordained for Seleucia-Ctesiphon, where he could directly represent the interests of his community to the civil government. From the time of the local synod of 315 the bishop of Seleucia-Ctesiphon was recognized as the head of the Church. This marked a decisive step towards the self-definition of the Church of the East. In consequence, the capital Seleucia-Ctesiphon was officially considered the principal see of the Church of the East.

The Church of the East in Persia under the Parthian (247 BC–224 AD) and Sasanian (224–637) Empires

To return to general history and geography, at the beginning of Christianity the frontiers of the Parthian Empire (247 BC–224 AD) extended from modern Afghanistan and the River Indus (today in Pakistan) in the east to the east bank of the Euphrates in the west, and from the Armenian frontier in the north to the south shore of the Caspian and to the Persian Gulf.

In 224 the Parthian Empire was overthrown by Ardashir I (224–40), the founder of the Sasanian dynasty (224–637). (The Sasanids were another Persian tribe.) The Sasanian Empire attained its greatest expansion under Khosroes II (591–628), extending from Armenia and Georgia in the west (the modern Caucasus) as far as modern Pakistan in the east, and comprising also territories in Syria, Egypt and Central Asia. The Sasanian Empire was recognized at that time as one of the main world powers, along with its neighbour, the Roman (Byzantine) Empire, between whom wars were regularly waged for a period of more than four hundred years.

Upon the conversion to Christianity of the emperor Constantine (306–37) and the signing of the edict of Milan in 313, legal and official status was given to Christianity in the Roman Empire. Constantine established his capital at Constantinople, laying the foundations for the East Roman, or Byzantine, Empire (330–1453). He was the first Roman emperor to be converted to Christianity. In Persia the state's official religion was Zoroastrianism.

This situation led to the widespread religious persecution of Christians (bishops, priests, deacons and laypeople) in Persia from 338 to 363, then more sporadically until 379. These persecutions were driven, at least in part, by distrust of the loyalty of Persian Christians vis-à-vis 'Western Christianity', that is to say, the Christianity of the Byzantine Empire, and so fear of potential treason to the benefit of the enemy power. This kind of accusation is frequently found in the historical sources.

In Persia some of the new religion's converts came from Zoroastrianism. For example, one famous patriarch of the Church of the East, Aba

I (540–52), had been the secretary of the governor of the province of Beth Garmai before his conversion to Christianity. For all these reasons, a certain number of Persian Christians were put to death, as one can read in the *Acts of the Persian Martyrs*.

Persecutions

The first persecutions of Christians began in the Parthian Empire. Before the reign of Shapur II (309–79) one cannot speak of organized and systematic persecutions, even if under Bahram II (276–93) Christians experienced difficulties. In the reign of Shapur II a widespread persecution took place which lasted for forty years. Persecution was revived sporadically under his successors. Persecutions also took place under Yazdgard I (399–420), Bahram V (420–38) and Yazdgard II (438–57) and until the reign of Khosroes II (590–628). Under Yazdgard II there was intermittent persecution, mainly in the Beth Garmai. Under Khosroes I (531–79) the resumption of hostilities against the Byzantine Empire led to renewed persecution.

Despite all this, the Church of the East continued to prosper. On the whole, it was tolerated by the Sasanian kings, even though, as has been explained, it was considered suspect during periods of war with the Byzantine Empire. Nevertheless, by the last decade of the fourth century the Church of the East had recovered.

The Autocephalous Organization of the Church of the East

At the beginning of the fourth century the Church of the East was well established in the capital, Seleucia-Ctesiphon. In the fifth century, particularly at the time of the synods of 410 and 424, the Church of the East was reorganized and the primacy of the bishop of Seleucia-Ctesiphon became well established. In 410 a synod was held in Seleucia-Ctesiphon which marked a major stage in the history of the Church of the East. This was the Synod of Mar Isaac, named after the metropolitan of that name of Seleucia-Ctesiphon from 399 to 410. In 410 it was declared that the metropolitan of Seleucia-Ctesiphon was the primate of the Sasanian Church of Persia. During this synod the doctrine of the First Ecumenical Council of Nicaea (325) was accepted formally by the Church of the East, and rules were laid down for the internal affairs of the Church in a series of canons. When the title of patriarch-catholicos was adopted at the end of the fifth century for the head of the Church of the East, the autonomous status of this Church in the Sasanian Empire and beyond was further emphasized.

Principal Dioceses and Metropolitan Provinces of the Church of the East in the Fourth and Fifth Centuries

In the Sasanian Empire the Christians organized parish communities which developed with the bishop at their head. In 225 there were about a dozen bishoprics which only existed for a maximum of one or two generations.

In the first quarter of the fourth century the Church of the East consisted of fifteen to twenty dioceses in the regions of Beth Garmai, Adiabene, Beth Aramaye, Elam and Mesene in Mesopotamia and in the east in the province of Khorasan. A number of dioceses seem to have been founded before the Sasanian period, some of which even go back to before the end of the second century, such as the dioceses of Kashkar in Beth Aramaye, Beth Lapat and Susa in Beth Huzaye/Elam, Erbil in Adiabene, Prath of Mesene in Mesene and also Karka of Beth Slokh (Kirkuk) in Beth Garmai.

In 410 the Church consisted of five metropolitan provinces with a little under fifty dioceses in Mesopotamia, Persia, Arabia and India. In geographical terms these metropolitan provinces were situated in the following regions: Nisibis, Adiabene, Beth Garmai, Mesene and Beth Huzaye/Elam.

Each metropolitan had suffragan bishops under his authority. In 420 Fars was also recognized as a metropolitan province, as was Beth Aramaye in 420. Almost all these provinces lasted until the end of the thirteenth century.

It has been established that there was a significant number of Christians in Persia, Mesopotamia and Khorasan, and also on the Arabian side of the Persian Gulf and on the neighbouring islands. From time to time bishops were also sent to India.

In the north, the bishop of Nisibis was enthroned as metropolitan of Nisibis and Beth Arabaye in 410, with a territory extending along both banks of the Tigris between Nisibis and Mosul and delimited by the cities of Shigar and Balad to the south and Lake Van to the north. What do we know about the maximum extension of the Church of the East to the north-west? In 424, at the Synod of Dadisho, the presence is noted of a bishop of the Church of the East coming from Armenia, whose name was Artashar. His diocesan centre was probably situated in the city of Akhlat (on the western shore of Lake Van). This diocese appears to have been assigned to the metropolitan of Nisibis a little after the Arab conquest (seventh century). To the south, in the province of Mesene, a significant number of Christians is reckoned to have lived under Sasanian rule. In 410 a metropolitan is found in this ecclesiastical province (today the region of Basra), situated on the edge of the great commercial port of Spasinou Charax, founded by Alexander the Great and the point of departure for India (an important centre of trade in spices and other merchandise).

In the same year (410), the bishopric of Erbil became the metropolitanate of Adiabene, which comprised the regions between the two rivers of the Great and the Lesser Zab (tributaries of the Tigris) and also the plains of Mosul and Urmia as well as the modern regions of Amadiya and Hakkari (today in south-east Turkey). The bishop of Karka of Beth Slokh (Kirkuk) became metropolitan of Beth Garmai (a region between the Lesser Zab and the River Diyala and bordered by the Tigris to the south-west). In the centre of Persia some dioceses appear in the fifth century, with the presence of bishops mentioned around 410 at Hamadan and Hulwan (this is the metropolitan province of Hulwan, called Beth Lashpar by the Christians of the Church of the East) and also at Ray (Beth Raziqaye). In the fifth century there was also a diocese to the south-east of the Caspian Sea, in the district of Gorgan. Even further east, at Khorasan and at Segestan, which then belonged to the Sasanian Empire, there were at least four dioceses in the same period.

The Synod of Dadisho (424)

In 424, at a council convened at Markabta, the Church of the East declared its autocephaly. This council is also known as the Synod of Dadisho (421–56), the name of the metropolitan at the time. The council declared that the metropolitan of Seleucia-Ctesiphon was the sole head of the Church of Persia: no other ecclesiastical authority could be recognized as superior to him, and no appeal to the bishops of the West could be admitted. The Church of the East thus distanced itself deliberately from the Churches of the Roman Empire and emphasized its autonomy as the sole Church of the Sasanian Empire. The acts of this council were approved by the Sasanian king Bahram V (420–38), who confirmed the freedom of worship accorded to the Assyrian Christians by his father, Yazdgard I (399–420), who himself had recognized the institutional legitimacy of the Church of the East within the state in 410.

In 484 at the Synod of Beth Lapat the Church of the East did not recognize the Council of Ephesus. In this period, for the majority of Christians living to the west of Persia, the Church of the East was considered as having fallen into heresy and schism. Why? How?

Theology: The Christology of the Church of the East

Until 431 there was no division within the Church. The theological and Christological debates of the fifth century provoked painful separations which have endured to our own day. The first division occurred in 431 with the Council of Ephesus, which deposed Nestorius, patriarch of Constantinople from 428 to 431. Nestorius (386–451) was condemned specifically because he called the Virgin Mary *Christotokos* (Mother of Christ) and not *Theotokos* (Mother of God, or she who has given birth to God) as had been proclaimed at the Council of Ephesus in 431. The Church of the East preferred to call the Virgin Mary *Christokos*, a formulation that was acceptable to it, seeing that this Church believes, as do the Churches which accepted the Council of Ephesus, that Christ is fully God and fully Man without division or confusion.

In the history of the Church, the Church of the East was accused of being Nestorian, that is to say, of following the doctrine of Nestorius. The Church of the East was at the time considered to be marginal and heretical. In fact, the Church of the East rejects this appellation and the qualifier, used in our own day, of 'Nestorian'. This qualifier is an erroneous inference, as will be explained below.

The Church of the East accepts the two natures of Christ (divine and human) as proclaimed at the Council of Chalcedon in 451: two natures (*physeis* in Greek, *physis* in the singular/*kyane* in Syriac, *kyana* in the singular) in one person (*parsopa* in Syriac, *prosopon* in Greek). Moreover, the Church of the East does not speak only of 'two natures' to express the divinity and the humanity in Christ, but also of two *qnome* (plural of the Syriac word which is *qnoma* in the singular). This was the second accusation made against the Church of the East: its formulation of 'two persons' in Christ, because this contradicted the Greek formulation 'one Person' (*hypostasis* in Greek) chosen at the council of Chalcedon. It is important to understand what the Church of the East really means to signify by the words 'two *qnome*', a formulation often translated incorrectly as 'two persons', which was taken to be heretical and provoked rejection by the other Churches. How then should one understand and correctly translate the Syriac word *qnome*? As the distinguished Syriac scholar Sebastian Brock explains, in the Church of the East *qnome* does not signify two persons (*hypostaseis* in Greek) but 'a property specific to each of the two natures individually' (or 'a set of characteristics') more than something individual. They would speak of two natures with their *qnome* and of a single 'subject' of all of Christ's actions. In an article entitled 'The "Nestorian" Church: A Lamentable Misnomer' Sebastian Brock explains this in detail (see below, bibliography to Chapter 1, theology section).

Moreover, the Syriac expressions habitually used to translate the key terms of this theological debate did not always have the same meaning as the equivalent words in Greek (the language at the time of the ecumenical councils). In consequence, these Syriac technical terms were not always used or understood in the same fashion by Christians speaking Greek, Latin, or Syriac. In each case, the use of the word *qnome* gave rise to serious misunderstanding. Nevertheless in our own time, following thorough studies carried out by specialist theologians and linguists on this matter, one can speak of a terminological misunderstanding in what pertains to the Christology of the Church of the East. As a matter of fact, the Church of the East and the Roman Catholic Church have understood this, as was expressed during their bilateral dialogue when, in 1994, Pope John Paul II and the catholicos-patriarch Mar Dinkha IV signed a common Declaration confirming that their Christological expressions were not grounds for heresy and separation. The two Churches on that occasion together accepted the term *Christotokos*, and did so to resolve the separation created since the Council of Ephesus in 431 (<https://news.assyrianchurch.org/common-statement-on-sacramental-life/>). It is a fact that the Church of the East does not believe that Christ is divided into two Sons (the Son of God and the Son of the Virgin Mary), which for some people is still the principal theological/Christological objection. The Church of the East confirms this denial. Furthermore, historical research has demonstrated that the theological differences outlined above are not the only reasons for the conflict that existed at the time between Cyril of Alexandria and Nestorius. There were other factors of a personal, political and cultural nature that also played a decisive role (see John A. McGuckin's article, 'Nestorius and the Political Factions of Fifth-Century Byzantium', listed below in the bibliography to Chapter 1, theology section).

The Theological Schools

In the history of the Church of the East, a fundamental role has been played by two theological schools in particular: that of Edessa (today

Urfa, in south-east Turkey) and that of Nisibis (today Nusaybin, also in south-east Turkey). The most important of the disciplines taught there was naturally the study of the Bible and its interpretation. The first theological school was founded by the bishop of the city, James, who attended the Council of Nicaea in 325. Nisibis was then in Roman territory but came under Persian control in 363. Ephrem the Syrian moved the school, also known as the 'Persian School' on account of the numerous students coming from Persia, from Nisibis to Edessa.

As already mentioned, the fifth century was the age of the theological controversies centred on the subject of Christology. In 489 the Byzantine emperor Zeno ordered the closure of the school of Edessa and expelled all the Christians of the Church of the East from the Byzantine Empire. The teachers and pupils of the Church of the East withdrew to a new school at Nisibis (then in the Persian Empire, 223 km to the east) founded by Narsai. At Nisibis one of the best known professors was the Catholicos Aba I (540–52) already mentioned, a great scholar who had himself studied at this school. In the second half of the sixth century most of the bishops and metropolitans of the Church of the East, including some of the patriarchs, such as Ishoyahb I (585–95) and Sabrisho I (596–604), were alumni of the School of Nisibis.

There were, of course, other schools of theology of the Church of the East where Syriac doctrine and culture were taught and transmitted. Some of these schools were famous, for example, those at Seleucia, Gundishapur, Beth Qatraye (today Qatar) and later Baghdad. There were also theological schools in numerous episcopal towns and in some of the great monasteries. These theological schools and monasteries were centres of religious as well as secular culture.

The Monasteries

The monasteries also played a major cultural and spiritual role in the ascetical history of the Church of the East. The monastic life required not only celibacy, simplicity with a vow of poverty, prayer and manual work, but also study. The monks studied, in the first place, the texts of the Bible as well as writings on the monastic and ascetic life. Among the writers of the Church of the East a number of monks have left literary treasures in Syriac and Arabic.

As in the tradition of Eastern Christianity as a whole, the monasteries were the seed-beds which produced almost all of the Church leaders, since the patriarchs and bishops are chosen, as a rule, from among the monks. The monks played a decisive role in the Church of the East for the organization of educational work and for the expansion of the Church of the East's missionary work. Thus the patriarch Timothy I (780–823) generally chose monks of the monastery of Beth Abhe to become metropolitans of the ecclesiastical provinces described as 'exterior'. We know of monasteries of the Church of the East not only in Mesopotamia, but also in Central Asia and even in China in certain periods.

The following are the names of some monastic founders and their very famous monasteries. Above the city of Nisibis, on Mount Izla, two great monastic figures are known, first of all Augin, or Eugene (fourth century), and then Abraham of Kashkar (sixth century). The monastery called Mar Augin was founded by an Egyptian monk, Augin, who established the first cenobitic monastery, that is to say, a monastery with a community Rule, with a group of his monastic disciples. He may be considered the founder of monasticism in Mesopotamia and Asia. The community on Mount Izla developed rapidly, and from there the monks established other monasteries situated mostly in the four northern metropolitan provinces of Mesopotamia. In 571 Abraham the Great of Kashkar (d. 588) founded and governed a new monastery on Mount Izla; he had numerous disciples. He and his successor, Babai the Great (d. 628), reorganized the monastic life in a strict manner. The monks of this monastery themselves became founders in turn of other monasteries. Numerous new foundations are attested in this period. In 1271 the monastery of Mar Augin was restored by the local bishop, Abdisho of Gaslona. Around the nineteenth century the Syrian Orthodox appropriated the monastery, which by then had been abandoned. This kind of transfer from one Church to another also took place with regard to certain monasteries on Mount Izla, such as that of Mar Abraham of Kashkar (in the nineteenth century), or that of Mar Yohannan.

Another very famous monastery of the Church of the East is that of Marga, situated in the metropolitan province of Adiabene and founded by Thomas of Marga (ninth century), who is well known for his book on the history of his Church, translated into English in 1893 by E. A. Wallis Budge under the title, *The Book of Governors: The Historia Monastica of Thomas, Bishop of Marga*. He had become a monk at the celebrated monastery of Beth Abhe (in Syriac, 'the house of the forest') situated near the Great Zab about 80 km to the north-east of Nineveh (Mosul) and founded by Rabban Yaqob of Lashom in about 595. We know that the diocese of Marga, which is attested between the eighth and the fourteenth centuries, included a large number of monasteries in the environs of the town of Aqra (to the north-east of Mosul).

With regard to the great monastery of Rabban Hormizd (to the north of Mosul, 2 km from Alqosh), it is very famous because it became the residence of one of the patriarchal lines of the Church of the East, as will be explained below. It is the monastery that remained active the longest among all the monasteries of the Church of the East, in view of the fact that it was founded by Rabban Hormizd the Persian in the first half of the seventh century. Since 1818 the monastery has been occupied by the Chaldean Catholics. It should be borne in mind that from 1516 the Mosul region was part of the Ottoman Empire. To the north-east of Mosul the great monastery of Mar Mattai became the most important centre of the Syrian Orthodox Church in northern Iraq. In the ninth century Metropolitan Isho'dnah of Basra, also an author, mentions the existence of more than a hundred monasteries of the Church of the East. Progressive Islamization gradually put an end to monastic life from the Seljuk period onwards, except for certain isolated monasteries. By the nineteenth century there were no longer any monasteries in the Church of the East.

The Arab Invasion (Seventh Century): Christians and Muslims

To return to the more general history, at the end of the sixth century relations between the patriarch Ishoyahb I (585–95) and the Persian king were excellent. It may be said that the patriarchate of Sabrisho I (596–604) marked a golden age of the Church of the East. At the beginning of the seventh century the Persians conquered several important cities in the Middle East: Antioch in 611, Damascus in 613, Jerusalem in 614 and Alexandria in 617. During this period a certain number of churches were established by the Church of the East in the cities of these regions captured by the Persians.

In the second quarter of the seventh century Islam, which originated in the city of Mecca (today in Saudi Arabia), emerged from the confines of the Arabian Peninsula and began to spread northwards into the Middle East. First of all, in 635 and 636, the Arabs invaded Palestine and Syria and then the neighbouring regions, including Egypt (in 639–42), at that time under Byzantine rule. In 639–40 they also took possession of the northern Mesopotamian region, but the city of Mosul remained a bastion of the Christians of the Church of the East. In 643, after the Muslims had penetrated its frontiers, the Sasanian Empire was reduced to nothing within the space of a generation. In 637 the Arabs took possession of the capital, Seleucia-Ctesiphon, and the seat of the Church of the East fell in turn. In 651 the last Sasanian king, Yazdgard III, was killed. In consequence, the Sasanian territories became the possession of the Arab Empire (or Caliphate). In the eighth century the latter extended from the Mediterranean and the Red Sea to the Oxus and the Indus, and from the Indian Ocean to the Caspian Sea and the Caucasus. This conquest by the Arabs of the territories of Syria and Persia resulted in the great majority of Christians of the Church of the East also coming under Muslim domination. In this period the Church of the East comprised ten metropolitan provinces and between sixty and seventy dioceses. The majority of its members neither fled nor converted to Islam. They were treated as a 'protected community' (*dhimmi*) obliged to pay the poll tax (*jizya*).

Paradoxically, with the arrival of Islam in the seventh century the missionary work of the Church of the East underwent an unexpected expansion in several ways and in different directions. In fact, the Arab conquest gave the Church of the East the possibility of penetrating west of Mesopotamia and entering towns formerly situated in Byzantine territory. For the first time these Christians of Persia were able to preach in Syria (with the creation of a diocese of their Church there in the seventh century), in Palestine (where there was a bishop of their Church in Jerusalem in the ninth century) and in Egypt (where there was a bishop in the eighth century). At the end of the seventh century important communities of the Church of the East were established in Syria and Palestine.

In the seventh and eighth centuries the Church of the East also created dioceses at Beth Aramaye and in the northern provinces, at Nisibis and Erbil. There are references to these in the book of Thomas of Marga already cited (*The Book of the Governors*). In the same period, the Church of the East also directed its attention to the eastern territories situated beyond the frontiers of the Persian Empire. As regards the number of Christians in the Church of the East, the peak came at the end of the Umayyad Caliphate (661–750).

Under Islam, with the passage of time the number of Christian conversions to the new religion increased and defections multiplied. The first discriminatory measures against the Christians occurred under the last Umayyads. But the Umayyad caliphs were in general more interested in taxing the Christians than in converting them. The Umayyad period was one of great stability. In 750 the Umayyad dynasty, whose capital was at Damascus, fell and a new dynasty, that of the Abbasid caliphs (750–1258), moved the capital in 762 from Damascus to Baghdad (about 36 km from Seleucia-Ctesiphon further up the Tigris). This explains why at around the same time, under the patriarch Khnanisho II (773–80), the seat of the Church of the East was likewise moved to Baghdad. It remained there for about five centuries. We know that the patriarch Khnanisho II was a very good administrator of his Church, as was his successor, the patriarch Timothy I (780–823). In Baghdad, as in Damascus, Christians could hold high office at the Arab courts.

During the eighth century the Church of the East tried to organize its presence in the vast eastern territories of Ray and Khorasan. The patriarch Sliba-Zkha (714–28) created metropolitan provinces for Herat and Samarkand and also for India and China. The patriarch Timothy I himself also created new metropolitan provinces. Under Timothy I the Church of the East further consolidated its presence in the West, in Syria and Palestine, with a metropolitanate at Damascus.

The Translation Movement under the First Abbasid Caliphs

After its foundation in 762 by the caliph Al-Mansur, Baghdad became a cosmopolitan capital. It was during the patriarchate of Timothy I (780–823) that Syriac-speaking Christians began to write in Arabic rather than Syriac. The first Abbasid caliphs began a movement of translating books, many of which were works dating from classical Greek antiquity, from Greek into Arabic via Syriac. These were books in the fields of philosophy, medicine and other scientific subjects (astronomy, physics and mathematics), for which they commissioned learned Christians, for the most part Syriac-speaking, as translators. These played a cultural role of immense importance for the transmission of knowledge in the recently constituted Arab world. It is thus that Greek sciences and Greek literature were transmitted to the Arabs and, moreover, in the Arabic language. This considerable labour had an effect on the intellectual history not only on the Muslim world but also on the Western European world as a result of translations made from the Arabic into Latin.

These translators belonged to different Churches (the Church of the East, and the Syrian Orthodox and Byzantine Greek Churches) but the scholars of the Church of the East were the most numerous. One of the most famous of these translators, Hunayn Ibn Ishaq (808–73), was a member of the Church of the East who came from Hirta (Al-Hira). He was an expert in philosophy and the sciences and was also the principal physician of the caliph Al-Mutawakkil (847–61). He was fluent in four languages: Arabic, Syriac, Greek and Persian. The works which he translated were mostly on medicine, astronomy and philosophy. He also wrote medical books himself, as well as a Syriac Lexicon.

This translation movement was promoted by the second Abbasid caliph, Al-Mansur (754–75), then by Al-Mahdi (775–85) and above all by Al-Mamun (813–33), who in 830 created in Baghdad an academy of translation and sciences called the 'House of Knowledge', also known as the 'House of Wisdom' (*Beth al-Hekma*), a centre of higher education that was part of the translation movement. In their translations of philosophical and apologetic works the Christians tried to defend their Christian faith. This was, for the times, a kind of inter-faith dialogue.

As had been the practice with the Persian kings and the Umayyads, the Abbasid caliphs surrounded themselves with Christians whom they employed as secretaries, administrators, accountants and physicians, a certain number of whom belonged to the Church of the East – men who often exercised considerable influence. Thus the caliph Ar-Radi (934–40) had an all-powerful secretary, Ibn Sangala. The majority of the physicians came from Gundishapur, one of the greatest medical and scientific centres of the medieval world, where there was a celebrated academy that taught medicine, philosophy and the sciences. In 709 the catholicos Timothy I tells how he financed the foundation of a hospital near his patriarchal residence in Baghdad.

During the Abbasid (750–1258) and Seljuk (1038–1307) Periods

In the age of Al-Muqtadir (908–32), at the time when the Chalcedonian Byzantine Greeks wanted to establish a prelature in Baghdad, only the catholicos of the Church of the East had the right to reside permanently in Baghdad; even the maphrian (the title of the primate of the Syrian Orthodox Church in the territory lying to the east of that of the patriarch of this Church) was not permitted to stay there permanently. At the end of the tenth century the patriarch Mari (987–99) was the first patriarch of the Church of the East to receive a letter of recognition from the caliph as a proof of his legitimacy.

During the Abbasid period there were about fifteen metropolitan provinces in the Church of the East. At the synod of 790 it was decided to create two classes of metropolitan provinces. Those called 'interior' (i.e. at the heart of the ecclesiastical territory, in Mesopotamia), and those called 'exterior' (meaning 'abroad', those of missionary activity at a distance from the centre). The metropolitans of the interior provinces were obliged to participate in the election of the patriarch and also had to report to him in person every four years. These interior provinces included the five traditional Mesopotamian provinces: Elam, Nisibis, Adiabene, Mesene and

Beth Garmai, and the adjoining province of Hulwan. The ten exterior provinces included Fars, Herat, and Merv and also other provinces created after 554: Ray (the city of Ray is situated today 10 km to the south of the city of Tehran), Armenia, Barda'a, Damascus, Samarkand, India and China. The metropolitans of the exterior provinces did not form part of the electoral college which could choose and elect the patriarch, but they had great autonomy and could ordain their own suffragan bishops. They were also required to submit a written report to the patriarch every six years, which facilitated the ecclesiastical administration. In this connection the long distances, which at that time did not always permit regular contact between the metropolitans and the patriarch, must be taken into consideration. According to the historian Eliya of Damascus, in 893 there were about eighty dioceses in the core territory of the Church of the East, the majority situated in Mesopotamia and Persia. At the beginning of the tenth century, the total population of the Church of the East, according to David Wilmshurst, numbered just over two million.

Confronted with Islam, the official religion of the Caliphate, the Christians sometimes experienced difficult times (persecutions, destruction of churches and monasteries), for example, under the caliphs Omar II (717–20) and Al-Mutawakkil (847–61). Under the Fatimid dynasty, the caliph Al-Hakim (996–1021) is well known for having conducted a long persecution against Christians and Jews in Egypt, Palestine and Syria, which lasted for ten years. These anti-Christian persecutions resulted in numerous apostasies, that is to say, conversions of Christians to Islam.

From the tenth to the twelfth century, in the provinces of southern Mesopotamia and in Persia, the Church of the East entered into a period of serious decline. In fact in the Middle East, according to Wilmshurst, there was a desperate period for the Church of the East from 906 to 1221. This did not impede the patriarch Sabrisho III (1064–72) from showing still more zeal in pursuing the expansion of his Church's missionary activities in Egypt, Jerusalem, Aleppo, Socotra and also in Chinese Turkestan. Egypt, after having been a simple diocese subordinate to the metropolitanate of Damascus, became a metropolitan province itself between the eighth and the eleventh century. Not much is known about the community of the Church of the East in Egypt, but it is known that at least two monasteries existed, one of which passed to the Copts in 1181 because there were no longer members of the Church of the East in that country. In 1099 Jerusalem was taken by the Crusaders. In 1187 Saladin captured Jerusalem, which seems to have provoked almost everywhere a wave of persecution against the Christians. An-Nasir (1180–1225), the thirty-fourth Abbasid caliph of Baghdad, promulgated a decree excluding Christians from official positions. Nevertheless, the following caliph, Az-Zahir (1225–6), was a just ruler, as was his son, Al-Mustansir (1226–42). Some Christians at that time were again appointed to important official posts.

During the Seljuk period (1038–1307) the metropolitan province of Nisibis still experienced some growth through the migration of Christians of the Church of the East from southern Persia (Beth Aramaye, Beth Huzaye/Elam and Mesene) who came to live there and in western Mesopotamia as far as the region of Mardin and Amid (Diyarbakir), towns situated today in south-east Turkey. In the south, the archdiocese of Basra (today in southern Iraq) is mentioned for the last time in 1222. In the tenth and eleventh centuries several monasteries were confiscated by the Muslims. The monastery of Beth Abhe remained an important seminary.

By contrast, in the eleventh century the Church of the East in Central Asia and Mongolia experienced spectacular growth. Although the Church's first missionary efforts that had begun in China in 635 had collapsed towards the middle of the ninth century, several large tribes dependent on the Mongols (1206–1368) converted to Christianity and became members of the Church of the East. In 1190 Yahballaha II (1190–1222), who was born in China, was enthroned as patriarch of the Church of the East. It was in his time that the patriarchal residence was transferred from Baghdad to Maragha, at that time the capital of the Mongol ilkhanate in Persia.

The Mongols (1206–1368), Timur Called Tamerlane (1370–1405) and Their Successors

In 1258, with the sack of Baghdad by the Mongols under the khan Hulagu (the grandson of Genghis Khan), the Abbasid caliphate came to an end. In fact, the period between the years 1222 and 1317 may be called 'the

Mongol century' in the history of the Church of the East. At the end of the thirteenth century, the Church of the East was still pursuing its missionary activity in the territories throughout Asia as far as China. The Church of the East reached its zenith in the thirteenth century and at the very beginning of the fourteenth century. At that time it consisted in total of twenty-seven metropolitanates, each comprising six to twelve dioceses, in a territory which extended in this period from the eastern Mediterranean to the Yellow Sea (China). Its missionary activities were focussed solely on the east. In the Middle East at the end of the thirteenth century the districts of northern Mesopotamia still constituted the heart of the Church of the East. In 1318, however, there began a period lasting until 1552 that may be called 'the dark years' or 'the dark night'. Although our information is scanty about these years, we shall try to summarize this Church's history.

Between 1330 and 1370 the Mongol Empire became fragmented. Then came the military campaigns of Timur Lang (Tamerlane) (d. 1405), marked by great violence and destruction in the regions through which his troops passed. That is why the feeble vestiges still persisting of the activities of the Church of the East were extinguished at that time in numerous places in Central Asia and elsewhere in Asia with the exception of India.

In southern Mesopotamia the surviving Christian communities disappeared between the fourteenth and sixteenth centuries. Christianity was also extinguished to the south of Mosul, in Beth Garmai. Following the massacre of 1310 at Erbil, laypeople and members of the clergy came to settle to the north and at Karamlesh (to the east of Mosul), which was then for a time the place of residence of their patriarch, who encouraged his flock to migrate towards the plain of Mosul in order to try to live in relative security. In consequence, in the middle of the fourteenth century the population of the Church of the East increased substantially in the Mosul region, at Karemlesh and just to the north at Telkepe and Tel Isqof.

With regard to the general history of the period, at the beginning of the sixteenth century the Ottoman Empire extended its territories eastwards and the Safavid dynasty became established in Persia. Wars were frequent between the two powers. The sixteenth century was a century of violence experienced regularly by the faithful of the Church of the East. The territories which they inhabited in Mesopotamia and Kurdistan often changed hands. In the middle of the sixteenth century the Church of the East's communities survived only in India and in the north of Mesopotamia, in a kind of triangle between Mosul, Lake Van and Lake Urmia, as well as a small number to the south and west of Mosul.

The Schism between the Church of the East and the Chaldean Catholic Church in 1553 and Its Aftermath

The seat of the patriarchate of the Church of the East was moved several times and to a number of different places, notably in consequence of changes of political regime. From Seleucia-Ctesiphon, it was moved to Baghdad as early as Timothy I's time (780–823) between the eighth and the thirteenth centuries. It was then moved north to Maragha (one of the capitals of the Mongol ilkhans) from 1268 and remained there under Patriarch Yahballaha III (1281–1317) moving to Erbil under Timothy II (1318–32), to Karemlesh under Denha II (1336/7–81/2), to Mosul from around 1385 to 1450, and to Gazarta (today Cizre, in south-east Turkey, to the north-west of Mosul) from about 1450 to 1504. Then the patriarchs resided at the monastery of Rabban Hormizd (51 km north of Mosul) from 1504 to 1804; this is the so-called line of Eliya (or Elijah). The patriarchal line of Shimun (1600–1918) settled at Kochanes (20 km to the north-east of the modern town of Hakkari/Julamerk in southeast Turkey). In the twentieth century the patriarchal seat was moved to Chicago (in the United States) and then in September 2015 to Erbil (in Iraq).

In 1450 the patriarch Shimun IV Basidi (c.1450–97) arranged that the patriarchal succession should stay within the family and become 'hereditary', passing in principle from uncle to nephew. This system should have permitted an assured patriarchal succession but it was to provoke serious conflicts at various times.

In 1552 the patriarch Shimun VII Ishoyahb (1539–58) of the Church of the East became unpopular and his opponents rebelled against him.

They then chose Yohannan (John) Sulaqa, the superior of the monastery of Rabban Hormizd as their ecclesiastical leader. In other words, Sulaqa was elected anti-patriarch by a faction within the Church of the East which disapproved of, among other things, the 'hereditary' principle of succession. The reasons for the schism as well as their consequences have been a point of controversy on both sides from that time up to the present.

The schism broke out in earnest in 1553 when Sulaqa went to Rome, where Pope Julius III enthroned him as 'patriarch of Mosul' thus making him the first patriarch of the Church known as the Chaldean Catholic Church. The latter then established himself at Amid (today Diyarbakir, in south-east Turkey). In 1553 the Ottoman Turkish authorities issued documents to him recognizing him as the head of the nation called Chaldean, but in 1555 he was arrested and put to death by the Ottoman governor of Amadiya. In 1681 another branch of the Catholic patriarchs was established at Amid. When this was suppressed in 1828, its dioceses were attached to those of the so-called Catholic patriarchate of Mosul. In 1830 Yohannan VIII (Hormizd) became patriarch of the Chaldean Catholic Church and the two seats were then reunited as a single one, installed at Mosul. This is the present line of the Chaldean Catholic patriarchs. In 1846 the Chaldean Catholic Church was recognized by the Ottoman Empire as a *millet*, that is to say, a distinct religious community. As for the Church of the East, it never received such recognition from the Ottoman government. It was represented to the government by the apostolic Armenian patriarch at Constantinople. So for a time the Chaldean Catholic Church was divided into two uniate administrations, that is to say, administrations united to Rome, at Amid and at Mosul. For a time there were also two parallel hierarchies in the Church of the East.

After the schism of 1553 the monastery of Rabban Hormizd remained within the jurisdiction of Patriarch Shimun VII Ishoyahb (1539–58) and his successors until 1804. The great majority of the faithful stayed loyal at that time to Patriarch Shimun VII. The monastery remained the seat of the patriarchs of the Church of the East throughout the seventeenth and eighteenth centuries. As early as 1497 the patriarch Shimun IV Basidi (1450–97) had been buried at this monastery and during the three centuries that followed all the patriarchs of the Church of the East except one were also buried there. Eliya XIII Ishoyahb (1778–1804) was the ninth and last patriarch of the Church of the East to be buried there. One can still read their epitaphs at the monastery.

Between 1558 and 1804 the history of the Church of the East is extremely complex because the allegiances of the patriarchs varied constantly. Some of the patriarchs of the Eliya line tried to established relations with Rome. It was under the patriarch Shimun X (1600–38), originally of the Church of the East from another patriarchal line, that the seat of the Church of the East was moved to the village of Kochanes in the remote mountains in the region of Hakkari (a region today in south-east Turkey). At the end of the seventeenth century there were two patriarchs of the Church of the East, one at Kochanes and the other at Rabban Hormizd. This situation lasted until Eliya III Ishoyahb died in 1804. Since the time of the patriarch Shimun XVII Abraham (1820–61), the patriarchate of the line of Kochanes (1600–1918) remained more or less uncontested. In order to give an account of these very complicated patriarchal lines, I have consulted the chronological table of the patriarchs of the Church of the East set out by David Wilmshurst in his book *The Martyred Church*.

Finally, let us recall the fact that the patriarch of the Church of the East Eliya XII Denha (1722–78), in particular, had to face a great challenge, that of the activities of Catholic missionaries. These missions received at the time strong reinforcement with the arrival at Mosul of the Dominicans, who had already been preceded by the Capuchins. Their proselytism was organized to apply pressure to villages inhabited by the faithful of the Church of the East around Mosul. All these vicissitudes of the Church of the East and the Chaldean Catholic Church, as well as their relations, with sometimes attempts to unite and then to disunite, can only be summarized here very briefly without attempting to explain them in detail.

Concluding Remarks

What we must emphasize in this history of the Church of the East is the fact that this Church has been among those which have experienced the greatest success in their missionary activities. In fact at the beginning of the fourteenth century the Church which had extended over the greatest area of the globe was not the Roman Catholic Church (then confined to Western Europe), or even the Byzantine Orthodox Church, but the Church of the East, with communities dispersed from Egypt to China and from Arabia and India to Lake Baikal. Although these feats were eclipsed in the course of succeeding centuries, they are of great interest for the history of the universal church, even if they are too often forgotten in our days.

During the patriarchate of Yahballaha III (1281–1317) the heart of the Church of the East lay in the plain of the Tigris to the north of Mosul, in the mountains of Bohtan and Hakkari and also in the region of Urmia (situated today in Iraq, Turkey and Iran): these are the regions in which the faithful of the Church of the East survived until the First World War (1914–18).

In 1348, the historian Amr spoke of twenty-seven metropolitan provinces between Jerusalem and China. In the mid-sixteenth century this structure had almost disappeared. There remained only India and the north of Mesopotamia. In 1552, the year of the schism, the Church of the East had no more than three bishops, at Erbil and in Persian Azerbaijan.

Following the schism of 1552 in the regions of Hakkari and Urmia, the faithful of the Church of the East remained loyal to the patriarchal line of Shimun which had fixed its patriarchal seat at the village of Kochanes. Most of the literature concerning the Church of the East was written for the most part before the fourteenth century, and the period between 1318 and 1914 is less well documented.

As has been too rapidly explained in this chapter, the establishment of the Church of the East in the Persian Empire gave it a free hand over large territories. There were at that time numerous great commercial routes which passed through Persia and Central Asia, such as the famous Silk Road, or the maritime routes that linked the Persian Gulf with India and beyond. Parallel to this, the Christianity of the Church of the East spread beyond the Persian frontiers, southwards to Arabia, south-eastwards to India and north-eastwards to Central and Eastern Asia along the Silk Road as far as China.

CHAPTER 2

In Arabia and the Persian Gulf

In the North of the Region of Arabia and in the Yemen

Before the Arab conquest (which began in 635), the missionary activity of the Church of the East was orientated in the following two directions: towards the south in the direction of modern Saudi Arabia (the birthplace of Islam) and the Yemen, and towards the south-east in the direction of the Persian Gulf (or Arabo-Persian Gulf) and India.

In Arabia the oldest Christian sources mention Christian missionary activities before the rise of Islam. The region of Arabia consisted of independent kingdoms, of which Hirta (its name in Syriac – Al-Hira in Arabic) lay on the direct route between Persia and Arabia. Hirta was also the ancient capital of the Arab tribe of the Lakhmids, among whom numerous notables in particular had converted to Christianity probably from the fourth century. The city lay to the south-west of Ctesiphon, the capital of the Persian Empire, in the south-west of Mesopotamia bordering the Great Arabian Desert (to the south-east of the modern town of Najaf in Iraq). The Lakhmids were vassal-affiliates of the Sassanid monarchs of Seleucia-Ctesiphon. The city of Hirta was also the seat of a diocese of the Church of the East whose bishop was also responsible for a certain number of Christian communities living in the south of Arabia and in the kingdom of Himyar, in what is today Yemen. To date, very little is known about the history of these Christians.

At the end of the sixth century (around 593), a bishop of the Church of the East, Shimun of Hirta, converted the Lakhmid king, Numan III, to Christianity. The Church of the East then became solidly established. It is known that the end of the diocese of Hirta came in the eleventh century. The ruins of Hirta can still be seen 3 km to the south of Kufa on the west bank of the Euphrates. At Hirta itself archaeologists have excavated two churches dating from the sixth and seventh centuries. The first excavations were conducted by Dr David Talbot Rice in 1931.

In its heyday Hirta could boast more than forty monasteries in its region, as is attested by archaeological sites. For example, as a result of the excavations made at Ain Sha'ia and in the caves of Dukakin situated not far away (15 km west of Najaf and 170 km to the south of Baghdad) archaeologists have brought to light one of the numerous monastic establishments dating from the eighth century, with traces of earlier foundations. At about 90 km to the north-west of Hirta, near Qusair, the ruins of two churches were found in 1931. In the course of later excavations Syriac inscriptions were discovered which prove the presence of Christians at these sites.

The city of Hirta (where Arabic was spoken but Syriac was also known) was an obligatory stop for merchants and travellers going to Arabia. From there Christians of the Church of the East were able to reach the very distant cities of Najran (today in the south of Saudi Arabia) and Sanaa (today the capital of Yemen) where other oriental Churches also had a presence.

In the Arabian Peninsula

At the end of the sixth century, the Sasanian army arrived in Aden and modern Yemen became a Persian province until the expulsion of the Persians by the Arabs in 630. It may reasonably be supposed that this explains how the Church of the East was able to profit from this new Persian conquest to assert its presence in these regions. We know that the bishop of the Church of the East at Sanaa around 840 was called Peter. The last known mention of Christians in these places dates from 911.

To the south of modern Yemen, the island of Socotra remained an isolated outpost of the Church of the East in the Arabian Sea. Here in about 525, according to the witness of Cosmas Indicopleustes (that is, 'Cosmas the Indian voyager', because he sailed to India), there was a large number of Christians of the Church of the East, with a bishop and priests appointed from Persia and dependent on Fars, which became a metropolitan province in 420. The ecclesiastical province of Fars was created around the port city of Rew Ardashir, on the Persian Gulf. Cosmas noted that there was a multitude of Christians at Socotra and that the diocese of this island was the third bastion of the Church of the East in the Arab world.

After the conquests of the Muslim Arabs, the population of Socotra, despite its ecclesiastical isolation, remained Christian. It is known that in 1281 the bishop of Socotra participated in the enthronization of the patriarch Yahballaha III. Around 1293–4, when Marco Polo visited the island of Socotra, he reported that all the inhabitants there were baptized and that their bishop depended on a patriarch who resided in Baghdad and not in Rome. In the following centuries, other important travellers reported similar information. On the two islands of Socotra and Ormuz (in the Persian Gulf and today in Iran), both of them situated in a zone of intense maritime commercial activity, it is reckoned that the communities of the Church of the East survived until about the beginning of the seventeenth century.

In the Persian Gulf

The ecclesiastical activities of the Church of the East were orientated equally in another direction, towards the south-east on both sides of the Persian Gulf because of its geographical proximity to the region of Fars (situated today on the south-west coast of Iran), a region where the Christian communities of the Church of the East had maintained a presence since about 225. According to Christelle and Florence Jullien, it is likely that the west coast of the Persian Gulf was also evangelized from Babylonia and the lands under the control of Hirta. It should be recalled that Fars was the heart of the Sasanian Empire and the cradle of Persian culture.

Since the second century BC, the Parthians had controlled the international trade from the Mediterranean that passed through Dura-Europos (today in Syria), then moved down the Euphrates as far as Spasinou Charax, an ancient port at the western point of the Persian Gulf (near the modern city of Basra in southern Iraq), and then went on to the Far East. This trade route continued to exist under the Sasanians (224–637) and under the Caliphates (from the seventh century). In consequence, the missionary activity of the Church of the East developed in a south-easterly direction from Mesopotamia and Persia using the riverine and maritime routes and passing along the Persian Gulf. Christian communities are found to have already come into being in the region of Basra in the second century, as is suggested by the *Acts of Thomas* (redacted at the beginning of the third century), with firmer evidence of a Christian presence in the third century. A propos of this evangelization by the Church of the East that was organized along the maritime routes via the Persian Gulf and as far as India, we should note that the Bishop David who went to evangelize India in around the year 300 was a bishop of Mesene.

With regard to the spread of Christianity in the Persian Gulf, it is possible that it began with the arrival of Christians who were fleeing the Persian persecutions at the end of the third century, or perhaps even before. In the islands of the Persian Gulf the presence of bishops is attested from the time of the council of 410, but there were some flourishing Christian communities before this date. Some Christians from Persia also came to establish themselves in the north of Arabia (that is to say, along the northern coast of modern Saudi Arabia and in what are today the Gulf States).

In the fifth century at least nine dioceses of the Church of the East existed in the region of Fars and along the Arabian coast of the Persian Gulf. At the beginning of the seventh century there were about a dozen dioceses in Fars and northern Arabia. The subjection of the region to Islam, which began in the seventh century, initiated the decline of Christianity together with the monastic life of the region. After the Arab conquest Christian communities continued to exist in the region of Fars for several centuries.

Three dioceses were grouped around the peninsula of Qatar under the Syriac name 'Beth Qatraye' (a zone sometimes called 'the Isles') which had their seats at the following places respectively: at Mashmahig (a town on the small island of Muharraq to the north-east of the island of Bahrein, in the kingdom of Bahrein), at Dairin (a town situated on the small island of Tarout, to the north-west of Bahrein) and at Hagar (near Tarout, a little way inland, in Bahrein). Further east, a fourth diocese, that of Beth Mazunaye, had its seat at the port city of Sohar also called Mazun by the

Persians (in modern Oman). As the dioceses on the north shore of Arabia were quite near the region of Fars, they were placed under the direction of the metropolitan of Rew Ardashir. At Qatar and at Oman it seems that Christians could still be found in small numbers until the tenth century.

This chapter in the history of Christianity is very little known to people outside specialist circles, but is quite well documented thanks to texts and archaeological excavations. On both sides of the Persian Gulf and on the islands archaeologists in recent decades have discovered the ruins of churches and monasteries of the Church of the East, a few examples of which will be discussed below.

Near Rew Ardashir, on Kharg Island (today in Iran) particularly important archaeological excavations were undertaken in the 1960s of a large monastery that brought to light sixty cells, a church, communal buildings (refectory, chapter house, libraries) and tombs, evidence of an ancient Christian presence. The island is now an important centre in Iran for the export of petroleum.

On the opposite shore of the Persian Gulf (also called today the Arabian Gulf) archaeological discoveries have been even more numerous. In Kuweit, on the island of Failaka, the remains of two churches were found together with a monastic complex at Al-Qasur. The remains of a church have also been found at Akkaz, a former island now part of the mainland. In Saudi Arabia, at Al-Jubail and also at Thaj (80 km to the south) churches, several hermitages and Christian graves have been discovered, and on the island of Dairin (today Tarut) also a monastery. In Bahrein, on the island of Muharraq, at the village of Samaheej, the old foundations of a monastery have been brought to light. In Qatar there are archaeological reports of the sites of at least two churches. In the emirate of Abu Dhabi, 170 km west of the capital, on the small island of Sir Bani Yas, archaeologists have discovered the ruins of a large monastery and a church at Al-Khor, and also another monastery on the island of Marwah. The majority of these Christian archaeological sites have been found on the coast or on the islands (in more remote and protected places, conducive to the contemplative life of monks), but also sometimes inland, for example, in the north of Saudi Arabia at Kilwa, where the remains of a church and monastic cells have been brought to light.

The existence of these Christian communities and the monastic life of the Persian Gulf is attested not only by archaeological discoveries but also by texts. In this region the dating of the buildings that have been found (churches and monasteries) remains very complex, as explained in the articles of Robert Carter (2008) and Jean-François Salles (2011), because either we have the presence of texts with the absence of archaeology (fourth to seventh centuries), or the absence of texts with the presence of archaeology (seventh to eleventh centuries). The first written mentions are in the acts of the council of 410; for the archaeology we have the survey of the oldest monastery of the Gulf, without doubt that of Al-Qusur, which according to Robert Carter dates from the second half of the seventh century. The dating of the archaeological material found in the Persian Gulf is based principally on the analysis of fragments of pottery. Robert Carter has 'redated' all the ceramics of the Christian sites of the Gulf to the eighth and ninth centuries, and perhaps to the ninth-tenth century for Kharg. All this remains to be examined in the future to attain greater precision. It has been established, however, that in the early years of the Abassid period (750–1258) there was quite a dynamic Christian presence in certain places of the Gulf region.

In Beth Qatraye, a little after the arrival of Islam in the first half of the seventh century, it has been ascertained that the Christian communities were still very prosperous. Beth Qatraye was an important centre for the Church of the East for its cultural and monastic life. In fact it is there that three of the greatest figures of this Church were born in the seventh century: Dadisho Qatraya, a celebrated spiritual author; Rabban Shapur, who established an important monastery in Beth Huzaye that bore his name and exercised an influence over the whole of Beth Qatraye; and Isaac of Nineveh (c. 640–700), a great ascetic and mystic of the Church of the East, who was born in Qatar and became bishop of Nineveh (today the city of Mosul). St Isaac is considered the most influential of all the writers in the Syriac language and he continues to exert a strong influence in numerous monastic circles in our own day. A number of other monastic and ascetical writers of the Church of the East were also born and brought up in the region of Qatar (Beth Qatraye). This reveals the presence of an important spiritual movement in the area.

In conclusion, both through a textual approach and through archaeology we have confirmation of the dynamic expansion of the Church of the East in the Persian Gulf and the tangible bonds which were maintained with the mother-Church in Mesopotamia. Throughout the Persian Gulf merchants and travellers, including Christians, could navigate from Spasinou Charax toward Fars, Arabia and, by crossing the Indian Ocean, as far as India.

CHAPTER 3

In India

From the Historical Beginnings to the Fifteenth Century

To reach the maritime routes of India/Southern Asia from the Middle East sailors had first to navigate the Persian Gulf, and from there, after crossing the Sea of Oman, make landfall on the west coast of India. From southern Mesopotamia (from the city of Basra, today in southern Iraq) to Kerala the voyage could last around three months and was undertaken in the sailing season when the winds were favourable for ships with sails. It was in Kerala, today a state in South India, that the community of the Church of the East developed, principally along the southern stretch of the west coast of India, which is also known as the Malabar Coast. Christianity also developed on the south-east coast of India towards Madras (today called Chennai, the capital of Tamil Nadu). Some think that in the sixth century there were also Christian colonies in Punjab, around the port of Tana (near Mumbai, formerly Bombay, on the north-west coast of India). The evangelisation of India is complex, with Christianity entering India in several different regions.

According to the Church of the East's tradition, it was St Thomas, one of Christ's twelve apostles, who was the first to evangelize Kerala from the time of his arrival in around 52, where he founded seven churches after landing at Cranganore. It is for this reason that the Christians of India are also called St Thomas Christians. In about 68, St Thomas met a martyr's death on the hill named after him at Mylapore (today a district in the south of Chennai, formerly Madras) where his tomb was still venerated in 1293, as witnessed by Marco Polo. In our own day Christians from India and elsewhere still go to the site on pilgrimage. Relics of St Thomas that were brought back from India to Edessa were also venerated. St Ephrem mentions them after his arrival in Edessa in about 363.

We know very little with certainty about the first centuries of the history of the Indian Church. Archaeological research has resulted in some discoveries, for example, metal plaques (in copper and other metals) and crosses of the Church of the East, some of them bearing inscriptions in the Pahlavi (or Middle Persian) language used in the time of the Sasanian Empire. The fact that such ancient objects have been found confirms that the presence of the Christians of India continued for the entire period up to the arrival of the Portuguese in India in 1498.

It is very difficult to give precise dates for the first contacts between the Church of the East in Persia and the first Christians of India. According to the *Chronicle of Seert* (ninth century), it was in 295 that David, the bishop of Mesene (in the Persian Gulf), left his country to be a missionary in India. The chronicle relates that a merchant, Thomas of Cana, came to India in the fourth century accompanied by a group of Persian Christians, amongst whom was a bishop called Joseph, perhaps originally from Edessa. They disembarked at Maliankara, near Cranganore. There are different versions of this story with different dates for the arrival of this Thomas in India; some speak of 345, others place it in the eighth or even in the ninth century.

In about the years 520–35, an Alexandrian merchant, Cosmas Indicopleustes, visited the coast of Malabar in south-west India and met some local Christians. He wrote a book entitled *Christian Topography* in which he speaks of Christian communities whose priests had been sent from Persia. He mentions a bishop ordained in the Fars region whose seat was at the Indian town of Kalliana (Quilon in Kerala), which confirms the link that existed with the Church of the East. This is the first precise and reliable historical source known to us. Cosmas also indicates the existence of churches in Ceylon (called Sri Lanka since 1972), where ancient 'Nestorian' crosses have also been found, which confirms the presence of Christians on this island for a long time.

In the middle of the seventh century some letters of the patriarch Ishoyahb III (649–59) mention that the line of bishops in India depended on the see of the metropolitan of Rew Ardashir in the Fars region (today in southern Iran). In the eighth century the patriarch Sliba-Zkha (714–28) seems to have reactivated the organization of the life of the Church of the East in the Indian province. After depending intermittently for several centuries on the metropolitan province of Fars, India became an autonomous metropolitan province under the patriarch Timothy I (780–823). This testifies to the growth of the Indian community at this period without our having precise references as to the number of bishoprics and where they were located. Patriarch Timothy I mentions the fact that some monks made a return journey to India.

We know of the existence in about 823 of another group that came from the Middle East to India led by Shapur and Peroz (perhaps two Syriac bishops) and accompanied by one Sabrisho (perhaps the metropolitan). Towards the end of the ninth century there is mention of a Metropolitan Yohannan who was enthroned for India. In general, we have no record of an unbroken succession of metropolitans resident in India. In periods of the *sede vacante* of a metropolitan or bishop, it seems that it was a local Indian archdeacon who then assumed ecclesiastical authority in a temporary manner. The patriarch Timothy I himself recognized the important role played by this archdeacon.

In the eleventh and twelfth centuries, during the Seljuk period (1038–1307), it was not always possible to send bishops to India and contacts with the patriarchate were sporadic. By contrast, at the very beginning of the fourteenth century contacts with the mother-Church were completely re-established thanks to the arrival of a new metropolitan. In 1301, under the patriarch Yahballaha III (1281–1317), we know that Mar Yaqob was metropolitan of India with his seat at Cranganore.

At the beginning of the fourteenth century there were still a few communities of the Church of the East around Bombay and on the Coromandel Coast (around Madras and Mylapore), in limited numbers. By the end of the fifteenth century these groups apparently no longer existed in north-west India. By the beginning of the sixteenth century these Christian communities had also disappeared on the east coast of India. The main groups lived along the Malabar coast where they have remained until the present day.

In 1490 two Malabar Christians made the long journey from India to Gazarta (today Cizre, in south-east Turkey) to the north-west of Mosul in order to ask the patriarch of the Church of the East, Shimun IV (*c.* 1450–97), to ordain a bishop for their Church in India. The patriarch ordained both of them to the priesthood. He also chose two monks of the monastery of St Augin (or St Eugene, today in south-east Turkey) and ordained them bishops. Then he sent them all to India, where new priests were also ordained. New churches were built at that time and the tomb of St Thomas was restored.

From the Arrival of the Portuguese in India in 1498

In 1498 the Portuguese set foot in India, at Calicut, under the leadership of Vasco da Gama. The discovery by da Gama of a new maritime route changed the course of world history and also that of India's first Christians. Before him, among the first Europeans known to have visited India and who mention the presence of these Christians of the Church of the East, may be mentioned John of Montecorvino in 1291, Marco Polo in 1293 and Oderic of Pordenone in around 1324. John of Montecorvino visited the tomb of St Thomas near Madras, as did Marco Polo, who reports that the fine earth that pilgrims collected from around the tomb had healing properties.

In 1503, thanks to Patriarch Elias V (1503–4), four bishops were sent to India, but only one survived the journey, Yaqob. The latter administered the Indian Church until his death in 1553. In his time the main group of Christians lived on the Malabar Coast, where they built yet more churches. In a letter written in 1504 it is specified that there were at that time about 1,400 churches for twenty or so towns in the region. The Portuguese estimated that there were between 100,000 and 200,000 families. Bishop Yaqob had to face serious difficulties in his relations with the Catholics. The competition that existed between the Church of the East and the Chaldean Catholics in the Middle East was thus also propagated in India.

In 1538 the first Latin bishop arrived in India at the town of Goa, which became the seat and centre of the Catholic mission in the country. This explains the conversions to the Catholic Church. From 1542 the Catholic missionaries showed their determination to convert the local Christians by force, wanting them to adopt the customs of the Roman Catholic Church by Latinizing their East Syriac rite and changing their local customs. In pursuit of this aim Joseph Sulaqa arrived in India. The brother of the patriarch John

Sulaqa, who had become a Catholic in 1553, Joseph was himself enthroned as metropolitan of India in 1555 by the Chaldean Catholic bishop, Elias Asmar. Joseph Sulaqa (d. 1569) lived in India from 1556 to 1558 and then from 1565 to 1568, for he was deported twice by the Portuguese. In fact, in 1558 he was arrested by them and sent to Lisbon, where he was interrogated by the Catholic Inquisition. Then he was given permission to return to India in 1565. But he was deported again, this time to Rome to be judged once more and it was there that he died. In 1559 a third Chaldean Catholic bishop, Abraham, after ordination by the Chaldean patriarch Abdisho IV (1555–70), was sent to India and established his see at Angamale (today a place near Cochin airport). On his arrival in India, he had to deal with the presence of another bishop, Shimun, who had been sent to India in 1558 by the patriarch of the Church of the East Elias VII (1558–91). On being denounced by Abraham to the Portuguese authorities, Shimun was arrested and deported to Rome, after which he was sent to live under house arrest in Portugal, where he died in 1559.

In 1597 the Catholic archbishop Aleixo de Menezes arrived in India. In 1598 he ordained more than a hundred Indian priests thus creating a Roman Catholic clergy parallel to that which already existed in India. In 1599 he convoked a synod at Diamper (Udayamperoor, to the south of Cochin), the aim of which was to uproot once and for all the Mesopotamian and 'Nestorian' influences in the faith, organization and liturgical life of the Indian Church. The converts had to promise to obey the pope and only receive bishops appointed by him, which subsequently provoked strong protests at the arrival of each bishop sent by the hierarchy of the Church of the East. At Diamper local Christians were required to renounce their Church of origin. Many were forced to become Roman Catholics, and in consequence to change their liturgical texts and use Latin as their liturgical language. Priests were sent to prison. Considered to be heretical, the books and manuscripts of the Church of the East were burned by the order of Archbishop de Menezes. This devastating work was continued by his successors. It was thus that the written cultural heritage of the Church of the East that existing in India in this period was destroyed.

By henceforth controlling the sea lanes in the Persian Gulf, the Portuguese were able to impede at will the arrival of new bishops and therefore any renewal of the hierarchy of the Church of the East in India. In doing so, the Portuguese Catholics succeeded in cutting off almost all contact between local Christians in India and their mother-Church in the Middle East. Communication between India and the patriarchate of the Church of the East became at all events difficult, not only because of the great distance that separated them.

Nevertheless, the patriarch of the Church of the East Elias IX Shimun (patriarch from 1617 to 1660) succeeded in sending a new bishop once again, Ahatallah, who arrived in India in 1652. But he too was arrested by the Portuguese because he was considered a schismatic. He was deported to Lisbon and delivered into the hands of the Inquisition; subsequently, according to some sources, he was put to death. In January 1653 a group of rebellious Indian Christians met at Mattancherry (Cochin). On the 16 January 1653 they signed a document called the Coonan Cross Oath. This was a public declaration affirming that the faithful of the ancient Indian Christian community of Kerala refused to submit to the domination of the Portuguese and the authority of the pope of the Roman Church in their lay and ecclesiastical life. The partisans of the Church of the East then sought a way to have another bishop. In 1653 twelve Indian priests organized the ordination of a new bishop there, called Thomas, but Thomas had to flee because he was threatened with arrest by the Catholics. All these events divided the Indian community into two parts: those who remained faithful to the Church of the East and those who submitted to Rome. Then a third Church made its appearance in India, as will now be explained.

In the course of their desperate efforts to have a non-Roman ecclesiastical hierarchy, some Indians after 1651 succeeded in coming into contact with other Oriental Churches in order to find a bishop who would not be Latin. That is how in 1665 a bishop of the Syrian Orthodox Church, Gregory (d. 1672) arrived in India. This accounts for the formation of a new Church in India, a Church of Western Syriac rite, which since 1912 has developed in two branches: the Church called today the Syrian Jacobite Malankara Church (which depends on the Syrian Orthodox patriarch of Antioch, with his seat at Damascus) and the Syrian Malankara Orthodox Church of India (autocephalous, with its seat at Kottayam).

In India

A group of Christians who rejected sharing the Western Syriac rite appealed to the patriarch of the Church of the East. A bishop called Mar Shimun was then sent to Kerala, but when he arrived in 1701 he was arrested and imprisoned by the Catholic Inquisition until his death. Other sporadic attempts were made to try and restore the bonds between the first Christians of India and their mother-Church, the Church of the East. In 1704 Patriarch Elias XI (1700–22) sent Metropolitan Gabriel of the diocese of Ardisha (Urmia), who promised to restore these contacts. He succeeded in gathering forty-two communities. But after his death in 1739 the Syrian Orthodox Church in India recovered them.

Nineteenth and Twentieth Centuries

In the nineteenth century the history of the Church of the East continues to be subject to disturbances, again because of dissension with the Catholic Church, both with the Latin-rite Catholic Church and the Chaldean Catholic Church of the East Syriac rite, both of them tied to Rome. Matters were complicated by the fact that certain prelates changed their Church by moving from the Church of the East to the Chaldean Catholic Church and vice versa. For example, Elias Mellus (1831–1908), born in Mardin (in the south-east of modern Turkey), was first enthronized in 1864 as bishop of Aqra (Iraq) by the Chaldean Catholic patriarch Joseph VI Audo, who then sent him to India. Then in 1874 he was reordained a bishop by the patriarch of the Church of the East, Shimun XVIII Rubil (1861–1903), who sent him back to India in order to assert his jurisdiction and serve the community in that country. Mellus was bishop at Thrissur from 1874 to 1882. In 1882 he was suspended and returned to Mosul. Then, after rejoining the Chaldean Catholic Church in 1890, he became in 1893 bishop of Mardin (in south-east Turkey). By contrast, a priest of Indian origin, in fact a Catholic, Antoine Thondatta, was ordained a bishop at Kochanes in 1862 by the patriarch of the Church of the East, Shimun XVIII Rubil. When he arrived at Travancore in 1863, an objection was raised to him there and he was forced to live as a simple Catholic priest. Then in 1882, under the name of Abdisho (d. 1900), he succeeded Mellus as metropolitan of the Church called the Syrian Chaldean Church (the name of the Church of the East in India) of Thrissur.

On the death of Abdisho in 1900, the community of Thrissur appealed to the patriarch of the Church of the East, Shimun XIX Benyamin (1903–18). The patriarch sent them Metropolitan Abimalek Timothy (1878–1945), born in the village of Mar Bishu (Turkey). He was enthronized in December 1907 and arrived in India in February 1908. Abimalek Timothy was head of the Church of the East in India until his death in 1945. It was he who organized the refounding of the Church of the East in India under the name 'the Chaldean Syrian Church', a name which may be confused with the Chaldean Catholic Church. But as a result of this, the ancient bonds with the Church of the East and its patriarchate were re-established. Three years after the arrival of Abimalek Timothy, a church group began a legal action against him. After fourteen years of judicial process, the final judgement allowed him in 1925 to recover the cathedral of Mart Mariam (of the Virgin Mary, consecrated in 1815) and all the properties of the Church. In May 2019 he was canonized by the Holy Synod at Erbil. This was celebrated with an official ceremony at Thrissur on 29 September 2019. His annual commemoration is to be on the 1 May.

It is thus that the present seat of the Church of the East was reinstated in India at Thrissur (the capital of the homonymous district, formerly called Trichur). In 1926 a publishing house was founded, Mar Narsai Press, which has been important for making liturgical books available, together with other books of the Church of the East's tradition in Syriac, Malayalam and English. After the death of Metropolitan Timothy in 1945, the diocese remained *sede vacante* for seven years.

It was Thoma Darmo, born in a village near Urmia (Iran) who succeeded Metropolitan Timothy as metropolitan of Thrissur from 1952 to 1968. Some disagreements broke out between Thoma Darmo and the patriarch Shimun XXI Eshai (1920–75). The patriarch suspended him from his functions by a letter issued in January 1964. In India the majority of the clergy and laity supported Thoma and refused to recognize the order of suspension, an act which led to a judicial process.

In India Mar Thoma had in June 1965 ordained Indian priests. In September 1968 he went to Baghdad, where he ordained three bishops: Mar Paulos (d. 1998) and Mar Aprem for India, and also Mar Addai Gewargis as archbishop of Baghdad. Then, in October 1968, thanks to the support of a group of members of the Church of the East in Iraq, Thoma Darmo (d. 1969) was enthronized in Baghdad as patriarch of the so-called 'Ancient Church of the East', an act which canonically raised the question of its true legitimacy, given that there was already an existing patriarch of the Church of the East, Shimun XXI Eshai (1920–75).

In September 1968, Mar Aprem, born George Mooken in Thrissur in 1940, became metropolitan of India with his seat at Thrissur, and Mar Paulos became his assistant. In 1971 the canonical patriarch, Shimun XXI Eshai, enthronized in Baghdad another metropolitan for India, Timothy (1920–2001), born at Thrissur. As a result there were two metropolitans of the tradition of the Church of the East in the diocese of southern India at Kerala. This was a very distressing schism. In 1976, the year of the abolition of the hereditary succession of the patriarchs of the Church of the East, a reconciliation was initiated between the two parties in India. In November 1995, after long negotiations, unity was finally re-established in India when Mar Aprem rejoined the New-Calendarist group of the patriarch of the East, Dinkha IV (1976–2015). Subsequently, Patriarch Dinkha IV made several official visits to India. It is to be hoped that this good example of reconciliation will be followed by the factions of the two patriarchates of the Church of the East that still exist in other countries throughout the world.

Concluding Remarks

As we have seen, in India, and particularly in Kerala, several Churches of the Syriac rite, including the Church of the East, have existed until our own day. At the time of Patriarch Yahballaha III (1281–1317), the name of the metropolitan of India was placed in the fifteenth position in the list of metropolitans of the Church of the East, without our knowing the seat at that time of the metropolitan in India. In the fourteenth century, southern India escaped the depredations of Tamerlane (1370–1405). On the other hand, under different conditions, the majority of the Christians of the Church of the East were gradually conquered by the Roman Catholic Church. In the middle of the sixteenth century the communities in India represented the sole external metropolitan province to survive in the Church of the East. In other words, with the exception of a small remnant living in the vicinity of the patriarchate situated at Kochanes (south-east Turkey), the sole community of the Church of the East that has managed to survive through the centuries to the present time – with a number of difficulties, it is true – is that of Kerala (in south-west India). It is very difficult to sum up the extremely complex history of the Church of the East in India and its tense relations with Catholic Church.

In 1806 an Anglican clergyman, Claudius Buchanan, visited the Kingdom of Travancore (which became the State of Kerala in 1956). He wrote: 'These Syriac churches on the sea-coast (of India) were compelled to acknowledge the supremacy of the pope in Rome. But they refused to pray in Latin and insisted on retaining their own language and liturgy. The churches in the interior would not yield to Rome, fled to the mountains and were compelled to seek protection of the native princes ...' A senior priest then told Buchanan, 'We are in a degenerate state compared with our forefathers .... The glory of our Church has passed away.'

In our own time, according to K. A. Ignatius, an author who belongs to the Church of the East and lives in Thrissur, historians are unanimous in saying that until the end of the fifteenth century, that is to say, until the arrival of the Portuguese in India in 1498 with Vasco da Gama and the Catholic missionaries, there was only a single Church, the Church of the East. This is an incontestable historical fact. He writes: 'Through its whole history it was an integral part of this Church under the supremacy of the Catholicos Patriarch of the East. How did the Church shrink into one of the smallest Churches in India? The greatest harm to the Church in India was done chiefly by the expansionist designs of and absorption by Roman Catholicism, with a process of annexation. This Catholic expansion and Syrian Orthodox penetration left the Church of the East in Malabar but an Assyrian remnant. One of the deadliest causes of the decline of the Church in Malabar was the absence of unbroken episcopal supervision.'

In India

The Syrian Chaldean Church in India constitutes at the present time one of the archbishoprics of the Church of the East, whose seat is still at Thrissur. It numbers about 25,000 members out of a total of Christians of Syriac rite in India estimated to be of seven and a half million. The Chaldean Catholic branch in India is called the Syro-Malabar Church. It is today the Syriac-rite Church with the greatest number of members, some four million.

CHAPTER 4

In Central Asia and Beyond. On the Silk Road

Geography of the Silk Road

Between the seventh and fourteenth centuries the Church of the East enjoyed a territorial expansion extending right across the continent of Asia, along the whole length of the Silk Road. The ancient Silk Road was a network of commercial routes that crossed Asia, principally Central Asia (modern Uzbekistan, Turkmenistan, Kazakhstan, Kyrgyzstan and Tajikistan) and beyond. All kinds of merchandise were transported along it, notably silk. The Silk Road was the principal commercial axis between Europe and Asia from the second century BC to the fourteenth century AD. It was along this road, from west to east, from the Mediterranean to China, on a trajectory of more than 10,000 km, that the Church of the East accomplished an extraordinary missionary expansion. The missionary monks of the Church of the East were the first Christians known to have reached Chang'an (today Xi'an), at that time the capital of China, in 635, during the patriarchate of Ishoyahb II (628–45).

In order to circumvent the very high mountains in southern Asia (the Himalayas and other chains), these missionary networks passed to the north of them, via the great commercial routes that went through staging-towns, some of which were oasis-towns, of Central Asia as far as China. In certain strategic commercial towns on the Silk Road episcopal sees were created, and then metropolitanates of provinces called 'exterior' in the history of the Church of the East. Thus, setting out from the west, where the Church of the East was centred, the missionaries were able to cross Central Asia by passing along the northern route going from Merv (today Mary in Turkmenistan) to Bukhara and then Samarkand (today in Uzbekistan) to arrive at Kashgar (today in the Uyghur autonomous region of Xinjiang, one of the five autonomous regions of the People's Republic of China). Another route, to the south, connected Merv with Kashgar via Bactres (today Balkh, in northern Afghanistan), then the capital of Bactria. To the east of Kashgar travellers could pass either by the north via the oasis of Turfan and Hami, or by the south via Yarkand, Khotan and Dunhuang. These two routes were separated by the basin of the River Tarim (in the south of Xinjiang), a great part of which is occupied by the Taklaman desert. Then the route continued to the east to arrive at Chang'an (Xi'an). These halts on the Silk Road are important for the history of the Church of the East between the seventh and fourteenth centuries (see <https://www.chinadiscovery.com/china-silk-road-tours/maps.html>).

This very cosmopolitan Road passed through the lands of people of different religions: not only Christianity but also Zoroastrianism, Manichaeism, Buddhism and Taoism, and also, after the Arab conquests of the seventh century, Islam. All were able to transmit their faith. The majority of Christians who used the Silk Road were merchants and missionaries, priests and monks. It was sufficient for them to follow the travellers who knew the routes that we described.

In Central Asia, from their region of Transoxiana (also known by the name 'Sogdiana', a name given by the Greeks to the territory situated today principally in Uzbekistan), Sogdian merchants through the centuries controlled the central part of the Silk Road in particular and areas well beyond it. They played a primary role for the transmission of religions along this axis, including the Christian religion, for the majority were members of the Church of the East. In fact, it was principally by their language, Sogdian, then the *lingua franca* of merchants along the Silk Road, that the Christianity of the Church of the East was mainly propagated. Later the Uyghurs played a similar role to that of the Sogdians.

It was in the sixth century that the Church of the East's missionaries penetrated Central Asia in a systematic manner. By the middle of the sixth century the influence of the Church of the East had gone beyond the frontiers of the Persian Sasanian Empire. In the first half of the seventh century its missionaries followed the Silk Road beyond the River Jaxartes (the former name of the Syr-Darya, today in Uzbekistan) as far as its eastern terminus at Chang'an, the capital of the Tang dynasty in China. Along the road they also converted Turkic and Turco-Mongolian tribes. In the eighth century the Church of the East's missionary activities in Central Asia had not yet reached the extent that we know of in the Seljuk (1038–1307) and Mongol (thirteenth to fourteenth century) periods.

History

We should remember that until the seventh century a large part of western Central Asia still belonged to the Sasanian Empire (224–651). This empire possessed territories in the provinces of Khorasan and, in the east, of Segestan. In these two former eastern Persian provinces (today straddling Iran and Afghanistan) there was a slow but nevertheless constant expansion of Christianity during the fifth century. To the east of Herat and Merv, in Bactria, the presence of Christians seems to be already documented in the third century. Thanks to the writer Bardaisan of Edessa, to whom the *Book of the Laws of the Countries* is attributed, we know that there were Christians to the south-west of the Caspian Sea and in Bactria in around the year 200. This is probably the oldest reference to Christianity in Central Asia. At the time of the Synod of Dadisho in 424, the presence of bishops of Segestan, Herat (today in western Afghanistan) and Merv is mentioned. At Herat the presence of a metropolitan in around 585 is known, and further north the city of Merv became the seat of a metropolitan in 544.

In the north of Bactria, along the eastern frontier of the Sasanian Empire and beyond the River Oxus (the former name of the Amu-Darya), a missionary expansion also began among the Hephtalites (called the White Huns by the Byzantines) and the Turkic peoples. A diocese was created for the Hephtalites in about 500, when their king made a request to the patriarch Aba I (540–52) to send them a bishop. By the end of the sixth century this Christian population was well established. When the Hephtalites (with territories to the north and south of the Amu-Darya) were conquered by the neighbouring Turkic tribes, who migrated westwards, a certain number of the latter also became Christians. In about 664, following an impressive miraculous event, Elias, the metropolitan of Merv, converted and baptized the king of a Turkic nomad tribe together with a large part of his people, in a region situated beyond Merv, to the east. Along the northern periphery of Transoxiana/Sogdiana and in the vicinity of the region of Khwarazm (to the south of the Aral Sea and the north of Merv, a region located today principally in modern Uzbekistan), other Turkic tribes were converted in their turn. By the advent of Islam (seventh century), Syriac Christianity was well established not only south of the Amu-Darya River (in Khorasan and Bactria), but also to the north of the river, among Sogdian and Turkic speakers in Sogdiana and adjacent areas.

The towns of Herat, and above all Merv, were strategic points of departure for missionary expansion in eastern Asia. Merv in particular (the capital of the Sasanian Persian province of Margiana, situated to the west of the Oxus) became a very important missionary centre because this town lay on a great crossroads of the Silk Road at the junction of the principal routes in the direction of China towards the east and in the direction of Seleucia-Ctesiphon and Antioch towards the west. There is no proof of the presence of a bishop or metropolitan at Merv before the time of the patriarch Isaac I (399–410). To the east of Merv, at Samarkand (Uzbekistan), the origins of Christian history still remain obscure. Some scholars think that the city was a key diocese in the mid-sixth century, but others are of the opinion that it became the seat of a bishop or metropolitan only at the time of the patriarch Sliba-Zkha (714–28), which would demonstrate the growth of the Church of the East in Transoxiana/Sogdiana in the seventh and eighth centuries.

The most important missionary expansion of the Church of the East took place with the encouragement of the patriarch Timothy I (780–823). Thomas of Marga confirms in his writings that this patriarch ordained bishops and then sent them to convert the non-Christians of the Far East. From his seat at Baghdad he organized new metropolitan provinces not only in the vicinity of the patriarchate but also at a great distance from it, as far as Central Asia and China. He sent well-prepared missionary monks there. In 792 or 793, Patriarch Timothy I, at the request of a local Turkic king recently converted to Christianity, appointed a metropolitan for Beth Turkaye, the land of the Turks.

In Central Asia and Beyond. On the Silk Road

At the dawn of the Arab conquests (seventh century), at Samarkand, Bukhara and Balkh, there were certainly communities of the Church of the East, even if there were not yet any bishops there. Kashgar (today in Xinjiang, in the north-west of China) was another crossroads city remarkably well situated to be a missionary centre of the Church of the East: beyond the high Pamir Mountains when approached from the west, at the foot of the Tian Shan Mountains lying to the north, and after crossing the Taklaman Desert when approached from the east. A bishopric was created at a date which is still uncertain, perhaps, according to some scholars, in the time of the patriarchate of Timothy I (780–823).

Starting from Merv and Samarkand and passing through Kashgar, the expansion of the Church of the East advanced further east, towards the Tarim Basin (today in north-west China), to the Turfan oasis and Hami in the north and to Khotan and Dunhuang in the south (<https://en.wikipedia.org/wiki/Tarim_Basin>). Then the route led eastwards as far as Chang'an. The presence of a Bishop John is known at Qamul, in the Hami oasis, in 1265.

With regard to the Uyghurs, there is currently no evidence for any Uyghurs converting to Christianity prior to their move to the Turfan Oasis in the mid-ninth century. We do know that the Uyghurs converted to Manichaeism in 762 CE, when Manichaeism became the state religion of the Uyghur Empire in what is now Mongolia, and indeed continued to be an important religion in the Uyghur Kingdom located in Turfan, as we know from the many Manichaean documents discovered there. After the move to Turfan, many Uyghurs also converted to Buddhism and a smaller number converted to Christianity, but both of these were gradual processes, probably beginning in the ninth century. Conversion to Islam did not take place until the fifteenth century.

The Uyghur territories extended from Almaliq (in the Ili basin, in the modern autonomous Kazakh prefecture of Ili in Xinjiang, <https://fr.wikipedia.org/wiki/Ili>) as far as Dunhuang, including the Turfan oasis. Dunhuang used to mark the demarcation between the Chinese Han and the Turkic populations. The River Ili rises in the Tian Shan (the Celestial Mountains) in the province of Xinjiang in China and flows into Lake Balkash in Kazakhstan.

In 1180, under the patriarch Elias III (1176–90), the number of Christians in the region of Kashgar and Nevaketh (to the north of Kashgar and south of Semirechye) was large enough for a metropolitanate to be established with ecclesiastical authority over the towns close to Tokmak (near Semirechye) and the region of Semirechye (the Russian version of the Turkic Zhetisu, the country of the 'Seven Rivers', situated today principally in the south-east of Kazakhstan, to the south of Lake Balkash, and also in Kyrgyzstan), another bastion of the Church of the East in central Asia. This is a further indication of the success of the Church of the East's mission in this region.

In the Tarim Basin (in the domains of Turkic speakers), at Kashgar, Khotan and Aksu (the latter being two towns to the east of Kashgar in the south and north of the Tarim Basin, today in the autonomous region of Xinjiang in China), it is known that there were Christians in the eleventh century – according to some scholars perhaps as early as the eighth century.

Moreover, around 795–8, Patriarch Timothy I announced his intention to enthronize a metropolitan for Tibet (Beth Tuptaye). In this period (eighth to ninth century) Tibet was an important independent kingdom whose territories extended as far as Central Asia and the town of Dunhuang (a place we shall come back to) which was under Tibetan control from about 781/7 to 848. At the beginning of the eleventh century, numerous Tanguts, who spoke a Sino-Tibetan language but belonged to a different ethnic group (not Turkic, like the Uyghurs), converted to the Christianity of the Church of the East in their new kingdom situated between the Tibetan mountains and Mongolia (<https://en.wikipedia.org/wiki/Western_Xia>).

During the eleventh century, a large number of Turkic and Turko-Mongolic tribes were converted, in turn, to Christianity in Mongol territory and adjacent areas. A letter of Abdisho, metropolitan of Merv, dated about 1007/8 and addressed to the patriarch Yohannan V (1000–11), mentions the conversion of about 200,000 'Turks' and their king. This episode of mass conversion is one of the most celebrated events in the history of the Church of the East. The story is told that, caught in a snow-storm, this king (*khan*) had a vision of St Sergius, who promised to save him if he would be baptized. Having been spared his life, the *khan* sent messengers to the metropolitan of Merv, who then assigned two priests to the king to baptize him and his people. According to the maphrian and Syrian Orthodox historian Bar Hebraeus, it would have been the Keraite tribe that was converted thanks to the metropolitan of Merv. Some historians therefore link this conversion with that of the Keraites, a nomadic group that lived in the region to the south of Lake Baikal in western Mongolia. Although some historians think that it was from the area inhabited by the Keraites that Christianity was able to spread to other neighbouring Turkic or Turko-Mongolic tribes, such as the Naimans, the Merkits, the Khitans and the Öngüts, others believe that the point of departure was not the Keraites but the Öngüts, or even the Oghuz.

It should be noted that the distinction between Turks and Mongols is not always clear. According to Mark Dickens, the Öngüts were a Turkic tribe. The Naimans spoke Turkish but were possibly of Mongolian origin and so could be called Turco-Mongols. It seems that the Keraites came from Tatars who spoke Mongolian but many of their names are Turkic and so they could also be called Turco-Mongols. The Merkits probably spoke a Mongolic language just like the Khitans. As for the Oirats, they were Mongols ethnically and linguistically. It was the governing élites of the Keraites and Öngüts who first converted to Christianity, followed by the mass conversion of their people. We know for sure that at least two tribes, the Keraites and the Öngüts, were already Christian before the Mongols came to power in 1206, for at that time they had Christian leaders.

Nomadic Tribes Converted to Christianity in Mongolia

It is therefore above all from the beginning of the thirteenth century, in the context of the Mongol conquests, that a remarkable expansion of the Church of the East may be seen in Central Asia and in the Mongolian region. In the middle of the thirteenth century the Keraites formed a solid Christian community. As already mentioned, the neighbouring tribe, the Naimans, also converted to Christianity, as did the Oirats, the Merkits and the Öngüts. In these regions of the nomadic tribes, the clergy had to follow their migrations. In fact, these tribes moved on horseback and in waggons and then used tents as places of worship. At the very beginning of the thirteenth century the Mongols brought all these tribes under their political control. They had no choice but to become allies of Genghis Khan (*c.* 1167–1227), the new Mongol sovereign.

These tribes lived in regions situated today in Mongolia, Inner Mongolia (China) and north-west China. It is difficult to place them precisely because they were nomadic tribes without fixed frontiers. According to Christopher Atwood (see his map in the *Encyclopedia of Mongolia*, <https://www.academia.edu/8855875/Encyclopedia_of_Mongolia_and_the_Mongol_Empire>, p. 390, and <https://en.wikipedia.org/wiki/Naimans>), the territory of the Keraites was in the zone lying to the north-west of the Gobi desert and to the south of Lake Baikal, with its centre in the Black Forest of the River Tula, a river that crosses central and northern Mongolia, and the modern capital of Ulan-Bator. According to some scholars, the head of the Keraites at the time of Genghis Khan occupied the country around Karakorum. To the north of the Keraites were the Merkits, who lived to the south-east of Lake Baikal (today situated in southern Siberia, in eastern Russia). The Genghis Khan's mother was a Merkit. To the south-west of Lake Baikal was the tribe of the Oirats. And still further west, near the Altai Mountains, were the Naimans. To the south of the Naimans was the Uyghur zone. Of the other tribes neighbouring the Keraites, to the south-east were the Öngüts, who lived in the south-east of the Gobi Desert near the great northern curve of the Yellow River and whose centre was the town of Olon-Süme (today an archaeological site of Inner Mongolia, in north-eastern China, 35 km north-east of the village of Bailingmiao, which lies to the north of Hohhot). It has been established that all these great nomad nations in Central and Eastern Asia in a large part became Christian. Among the Keraites and the Öngüts it is possible that there were still Christians until the middle of the fifteenth century.

The Tanguts (who lived to the east of Dunhuang, in the modern provinces of Ningxia and Gansu in north-west China) and their prince Nayan were also Christians. In fact in 1287, at the time of his revolt against the Great Khan of the Mongols, Khublai (1260–94), who had made himself emperor of China, we know that Nayan had the cross represented on the banners of his army.

In the thirteenth century monks, priests and prelates of the Church of the East were constantly travelling up and down the Silk Road. For example, in 1265 we learn that the metropolitan of Hami left his town (today situated in the eastern part of Chinese Xinjiang) to attend the enthronization of the patriarch Denha I (1265–81) in Baghdad.

In the fourteenth century the Church of the East was still flourishing along the Silk Road, as travellers from Europe and elsewhere testify. In their writings these travellers give us precious information. Thus, in his travels in eastern Asia (1272–95), Marco Polo testifies that he still saw prosperous communities of the Church of the East in the cities which he visited along the Silk Road, for example, at Samarkand, Kashgar and Yarkand. He mentions the name of a church at Samarkand dedicated to St John the Baptist, apparently constructed by the *ilkhan* Jagatai (d. 1242), the second son of Genghis Khan.

At the end of the Mongol conquests (thirteenth to fourteenth century) all the missionary activities of the Church of the East appear to have rapidly declined. It seems that numerous Christians were massacred and the monasteries and schools that depended on the Church were destroyed *en masse*. The exceptional missionary labours of the Church of the East, organized in the course of about a millennium in Central and East Asia were at that time reduced to nothing. In the fourteenth century the patriarch Denha II (1336/7–81/2) resided in the village of Karamlesh to the east of Mosul. It was in his time, in the second half of the century, that Christianity was eradicated in Central and East Asia. In Central Asia the Christians of the Church of the East are mentioned for the last time in the 1340s, except for Samarkand, where they still existed in the first half of the fifteenth century. In China the Christians were driven out following the overthrow of the Yuan Mongol dynasty in 1368. After the collapse of the Mongol Empire in Central Asia and China it is not known precisely how long the Church of the East's communities survived, perhaps for one or two generations. With regard to Central Asia, we still lack details about the history of the end of the Church of the East, which was brought about not only by the destruction due to the Mongols, but also by other causes: different kinds of persecution, epidemics, and also the isolation that these communities endured because of the great distances separating their metropolitans from their patriarch.

Languages and Translations

What languages did the missionaries of the Church of the East use? With regard to the liturgical language, not only Syriac was used but also other languages spoken by the people who were converted. These missionaries knew several foreign languages. As regards the written texts of the East Syriac tradition, to make them accessible to the faithful a number of translations were made from Syriac into local languages, principally Middle and New Persian, Sogdian, Old Turkic and Old Uyghur (similar languages), and Chinese. Sogdian became the second language after Syriac most used by the Church of the East in Central Asia. With the aim of teaching Christian life and doctrine, translations were made not only of liturgical texts but also of other texts, especially the New Testament, the Psalms, biblical and liturgical commentaries, homilies, texts of canon law, and also ascetical works treating of the monastic life, and the lives of martyrs and saints. We have proof of these translations thanks to the finding of ancient manuscripts, like the discovery made at the beginning of the twentieth century of a number of important manuscripts of the Church of the East in the Turfan Oasis and at Dunhuang, places that lay on the Silk Road.

Archaeology

The meagreness of written sources along the Silk Road, due principally to the fact that many of these Christians were nomads, has been progressively improved by fascinating new archaeological discoveries that allow us to fill out our study of the history of the Church of the East. The results of excavations published by scholars prove the presence of the Church of the East, and its expansion in Central Asia and beyond, before and after the Arab invasions. The inscriptions that have been found in the East Syriac language confirm that the finds relate to Christian communities belonging to the Church of the East. Here are some examples from places which are surprising to the non-specialist.

In Turkmenistan, near the ancient town of Merv (today Mary), archaeologists have found Christian graves (third to sixth/seventh century) in a cemetery, together with a church, a monastery and other religious buildings. In the south of Turkmenistan, at Ak-tepe, seals have been found, among other things, engraved with a cross in the tradition of the Church of the East. On seeing the remains of the church of Kharoba–Kochuk (even though the building has not been definitively established as such by some scholars), which lies to the north of Merv on the road that lends to Khwarazm, constructed probably in the fifth/sixth centuries and was perhaps active until the eleventh/twelfth centuries, one can imagine what the architecture of the Church of the East looked like in Central Asia.

In the whole region of Central Asia only five Christian buildings have been identified and excavated, of which the church of Urgut (39 km southeast of Samarkand) in Uzbekistan, probably constructed in the seventh century and occupied until the thirteenth century, is the most important archaeological discovery made between 2004 and 2007 relating to the Church of the East in Sogdiana. Also in Uzbekistan, near Urgut, Christian archaeological remains have been found at two other sites, at Koch-Tepe and at Quq-Tepe. At Afrasiab (the former site of Samarkand) a number of ossuaries with Christian symbols have been excavated that are not later than the seventh century. Thirty kilometres to the south of Samarkand, near Sufyon, a medieval Christian presence of around 752–3 is attested by inscriptions in Syriac. With regard to north-western Khwarazm (the ancient Chorasmia), the fact that Christianity had reached this region by the end of the seventh century is proved by ossuaries decorated with crosses in the tradition of the Church of the East that have been unearthed at Mizdakhan, to the east of the Caspian Sea. In the environs of Bukhara numerous coins have been found with crosses, most of them dating from the end of the seventh or beginning of the eighth century. Another important Christian community of Sogdiana in the Shash Oasis (Tashkent) is that of Qarshovul-Tepe, where a grave has been discovered and a bronze pendant in the form of a cross (<http://www.exploration-eurasia.com/inhalt_english/frameset_projekt_7.html>). At Tashkent gravestones may be seen bearing inscriptions in Syriac at the National History Museum of Uzbekistan.

In Kazakhstan objects of the Church of the East are displayed in the Central State Museum at Almaty. In this country numerous Christian tombs have been found bearing inscriptions in Syriac and Turkic, for example, at Taraz (ninth to tenth century). We know that churches were converted into mosques following the Arab invasions, for example, on the capture of Bukhara in 709, at Taraz (Talas) in about 893 and at Mirki before 985. A very important site was discovered in 2014: the ancient town of Ili-Baliq ('the town of the River Ili') in the village of Usharal, 53 km to the west of the ancient town of Almaty (the capital of the khanate of Jagatai, on the Silk Road) where a community of the Church of the East lived, as is attested by gravestones (*kayraks*) with inscriptions and engraved crosses (<http://www.exploration-eurasia.com/inhalt_english/frameset_projekt_aC.html>: see the maps). From 1400 to 1425, various tribes (Uzbek, Oirat, Kazakh, Kyrghyz, etc.) crossed the Ili valley and devastated it. The large medieval town of Ili-Baliq was one of the sites abandoned at that time. We know that there were Turco-Mongolic tribes in this region, the Naimans and the Keraites, especially to the east of Ili-Baliq, who became Christians in the eleventh century. The study of local coins and ceramics will allow us to say precisely when and why these Christians disappeared not only at Ili-Baliq but also in the whole region of Semirechye (today the district of Zhetysu, Almaty). In 2017 a centre for Nestorian studies was created at Almaty with the Archaeological Institute of the Academy of Sciences of Kazakhstan, where the Sixth International Conference 'Syriac Christianity in China and Central Asia' was held in June 2019 (<https://nestorianstudies.org/index.php/en/>). Since 2004 the Society for the Exploration of EurAsia, founded by Christoph Baumer, has been very active in the preservation of archaeological sites along the Silk Road (<http://www.exploration-eurasia.com/inhalt_english/frameset_projekt_aC.html>).

In Tajikistan, at Penjikent, some mural paintings (seventh to eighth century) have been discovered, a bronze cross and coins with the sign of the cross, and also a clay shard (an *ostracon*) dating from the end of the seventh or beginning of the eighth century on which the first two psalms are engraved from the Peshitta version of the Bible, the oldest translation into Syriac of the Old and New Testament.

In Kyrgyzstan near the village of Pokrova (close to Taraz) on the southern shore of Lake Issyk Kul Christian cemeteries have been brought to light with gravestones bearing inscriptions in Syriac and Turkic. At Krasnaya Rechka, near the ancient town of Nevaketh (in the valley of the River Chu, 400 km to the north of Kashgar and 15 km from Suyab), pectoral crosses have been found (eighth to tenth century) and an inscription in Sogdian which suggest that there was also the presence there of a community of the Church of the East. Moreover, silver dishes, probably made in Sogdiana (perhaps in Kyrgyzstan), dated to around the seventh-tenth centuries, one with representations of Christian scenes, including the Crucifixion and Resurrection of Christ, have been found in the region of Perm (in the Urals, at the western edge of Siberia).

Of the five churches discovered to date in Central Asia, two are located in the medieval town of Suyab (Ak-Beshim), another town on the ancient Silk Road, 56 km east of Bichkek, in the 'Seven Rivers' region (Semirechye). The most important archaeological remains found there, near Tokmak (in the villages of Burana and Kara-Jygash), include a monastic complex (tenth to eleventh century) at Ak-Beshim. Also found there were more than six hundred gravestones bearing inscriptions principally in Syriac (and also some in a Turkic dialect), but all written in the Syriac alphabet and dating mostly from the thirteenth and fourteenth centuries. The latest gravestone discovered in this region is dated 1345. It is the last known evidence of the presence of Christians in Central Asia in this period. Christianity in Central Asia declined rapidly in the fourteenth century. In the fifteenth century some sources still mention Christians, but only in Samarkand and possibly in Turfan.

In eastern Ladakh, which was part of the Tibetan Empire from about 644 to 842, near Tangtse (today in northern India) crosses have been discovered with an inscription in Sogdian (dated, according to various experts, between 825 and 842, (<http://www.shlama.be/shlama/content/view/214/180/>) which perhaps alludes to a monk of the Church of the East. Tangtse was situated on a transverse branch of the Silk Road which connected Sogdiana and Bactria with central Tibet.

All these excavations and archaeological studies provide us with reliable new information on the expansion of the Church of the East along the Silk Road, and give us a better idea of the history of the Christian communities that once existed in Central Asia.

The Coming of Islam to Central Asia

As already explained, the Church of the East experienced its first period of growth under the Sasanian Persians (224–651), then another period of rapid growth in the first two centuries after the Arab conquests, when the Church began its missions in Central Asia and China. After overthrowing the Sasanian Empire in 651, the Arabs crossed the River Oxus in 653–4. The Sogdian city-states sent embassies to Pekin to persuade the Chinese emperor to intervene, but without success. In 661 and 692 the Chinese reconquered the main part of the Tarim Basin and were recognized by the kings of Kashgar and Khotan. The Arabs attacked Sogdiana/Transoxiana in 657 and 683. The systematic Muslim conquest of Transoxiana began in 705. In the same year Qutayba Ibn Muslim (assassinated in 715) became the new governor-general of Khorasan and made Merv his capital. Sent by the Umayyad caliphs (661–750), whose capital was Damascus, he took Bukhara in 707, Samarkand in 712 and Tashkent in 714. In 722 Arab forces reached the banks of the Syr-Darya. In 750 the Chinese took Tashkent. Chinese and Arabs then found themselves face to face. In 751, at Talas in the valley of the Syr-Darya (today Taraz in Kazakhstan), the Chinese army was defeated by the Arabs. This very important battle marked the end of Chinese control in Central Asia in this period. As Édith and François-Bernard Huyghe write, after this event one could say that the Silk routes became the routes of Islam, and instead of becoming Chinese, Central Asia became Muslim.

It is thus in the eighth century that Islam began really to establish itself in Central Asia. At Samarkand the situation only became stable in the ninth century. The Muslim religion was by that time well established in the whole of Transoxiana. In the western region of the Tarim Basin, the inhabitants began to be converted to Islam in large numbers at the beginning of the khanate of the Karakhanids (840–1212). At the end of the fourteenth century the Uyghurs of the Turfan region were forced to convert by an inheritor of the Jaghatai khanate loyal to Tamerlane. By the sixteenth century, Islam had become the dominant religion in Central Asia and north-western China.

Wherever Islam was installed, the other religions disappeared, including Christianity. The Church of the East was also affected. The struggle between the Muslim Arabs and the Chinese (until 751) made communication more and more difficult for the Church of the East. The rise of Islam in Asia not only broke the links between the Church of the East in Persia and the communities in China and other places in Central Asia, but also crushed the infrastructure itself of the Church of the East in Asia.

Concluding Remarks

After the fall of the Mongol Empire in the fourteenth century the overland Silk Road never regained its former prosperity. To make the return journey to China merchants preferred to use the maritime route.

Once the Church of the East had converted the tribes of Central Asia and beyond, it could have been called the 'Church of Asia'. As we have seen, this Church had a presence in numerous places across Asia. Its missions along the Silk Road were among the most glorious achievements of the Church of the East and of the history of Christianity as a whole.

We conclude this chapter with some words on the Silk Road today. In the past, the Silk Road was known as 'the world's axis' because to control this route meant to 'possess' the world. In our own day the Chinese have reactivated a new Silk Road in order to link up with other routes in Asia, and also with Europe, one of their most important economic partners. In 2013 they inaugurated a motorway, together with a railway, whose point of departure is Chongqing, the largest metropolis of central west China today, lying 755 km south of Xi'an, the town where the missionary monks of the Church of the East arrived in 635.

CHAPTER 5

In China under the Tang (635–845) and the Mongols (1206–1368)

First Arrival in 635

Very far from the eastern frontiers of Persia and beyond the territories of the Turco-Mongol tribes of Central Asia, the missionaries of the Church of East, monks and priests, penetrated still further eastwards until they reached the great empire of China. In 635 they arrived at the capital Chang'an (today called Xi'an). Chang'an was then a cosmopolitan city situated at the crossroads of important routes and at the eastern terminal of the Silk Road. Indeed, the history of the Church of the East can be traced in two periods: under the Tang dynasty (in 635–845), and then during the time of the Mongol Yuan dynasty in China (in 1271–1368). These fascinating pages of Church history, even though often ignored by many Christians, reveal that missionaries of the Church of the East were the first Christians known to have arrived in China as early as the seventh century, almost a millennium before the organization of the first Catholic missions.

It is thanks to the stele of Xi'an that we know that a monk of the Church of the East, Alopen, arrived in China in 635, in the time of the patriarch Ishoyahb II (628–45) and during the reign of Taizong (626–49), the second emperor of the Tang dynasty (618–907). The empire of the Tang was at that time the greatest world power and maintained commercial relations as far as Byzantium in the west. The influence of the Tang extended into Mongolia, all the Central Asian steppes, the Tarim Basin, Sogdiana, Bactria, all the eastern part of modern Afghanistan, and also a northern part of India. The Tang era is considered a veritable golden age in Chinese history and culture. In the philosophical and religious domain, although Confucianism was the state ideology, it was Taoism that had the support of the first Tang emperors. As for Buddhism, which arrived in Chang'an about half a millennium before the official arrival of the Church of the East, it spread rapidly. Under the emperor Taizong, other religions also coexisted in China such as Zoroastrianism and Manichaeism.

The beginning of the history of Christianity in China is attested by the celebrated Tablet or Stele of Xi'an, the principal document that we possess, which is today exhibited in the Steles Museum or Beilin Museum at Xi'an. The Japanese scholar Pierre Yoshiro Saeki produced an English translation. This stone stele, 2.8 metres high, was erected in 781 on the land of the first monastery of the Church of the East founded in 638 near Chang'an by the missionary monk Alopen, who was most probably the leader of a group of missionaries. The stele was buried in 845 and then rediscovered and disinterred in 1625, also near Chang'an. These monks were probably not the first Christians to arrive in China, which is easy to imagine since we know that already in the seventh century there were numerous commercial links between Sasanian Persia and China.

During this first period, under the Tang, Christians formed only a very small religious minority consisting principally of people of Persian and Sogdian origin. The names of the monks inscribed on the stele of Xi'an reveal that some were of Persian or Syriac origin, but also that there were monks of other ethnic origin. The Chinese name 'Alopen' is probably a transcription of the Syriac name 'Yahballaha' meaning 'God has given'. Some of these first monks or Church members were most probably Chinese. We know that the inscription on the stele was composed by the metropolitan Adam of Beth Sinaye. Several bishops are mentioned: Yohannan, Yazdbuzid and Sargis (Sergius). There is also mention of priests and archdeacons, Gigoi of Khumdan at Chang'an and Gabriel of Sarag at Luoyang, the two Chinese capitals. It appears, then, that the hierarchy of the Church of the East was at that time well developed in China.

The first official contacts between the missionaries and the Chinese authorities apparently took place on their arrival in 635, when the monk Alopen was received at court in the imperial capital of Chang'an by a minister of the emperor Taizong. The Christian religion was recognized in an official manner in 638 by an imperial decree in which the emperor authorized the propagation of Christianity throughout the empire. The emperor also ordered a monastery to be built at his own expense at Yining Ward, near the capital, where twenty-one monks lived and probably also a bishop.

On the Xi'an Stele may be read not only a summary of the history of this first local Church in China, but also the main characteristics of the Christian faith. At the top of the stele there is the representation of a cross. The heading of the text, written in Chinese, announces the missionary aim: 'Monument Commemorating the Propagation in China of the Luminous Religion from Daqin', which gives the abridged name by which the stele is known: 'The Stele of the Luminous Religion'. Daqin is the ancient name which the Chinese used for the Eastern Roman Empire (Byzantium), and in particular, Syria (Antioch was the ancient terminus of the Silk Road, to the west of Persia). In China the Christian religion was consequently called Jing jiao, the Luminous Religion (literally, 'the luminous teaching', also translated as the 'venerable' or the 'illustrious' teaching). The text of the stele contains passages in Syriac, but is mostly written in Chinese, in such a way that the Chinese reader could comprehend or appreciate the message given. In order that the concepts and teachings of this new religion should be transmitted clearly to the Chinese, terms were also chosen that were adapted to local Chinese culture with a vocabulary simultaneously Taoist and Buddhist – a good example of inculturation. It may be said that this tablet is a masterpiece of intercultural communication. The stele is still undergoing study. As for the founding texts of Christianity (such as the New Testament), these were translated into Chinese in the imperial library, having first been examined by the emperor Taizong, who found them admirable, essential and useful to all.

According to David Wilmshurst, in the Tang period only about a dozen Christian colonies can be identified with churches and monasteries: along the Silk Road in the two capitals situated in the north of the country at Chang'an and further east at Luoyang, and also in the southern ports, numbering in total several thousand members. During the reign of Gaozong (649–83), Taizong's successor, Christianity spread across the ten provinces of the empire, according to the Xi'an Stele, which seems an exaggeration to some historians. Numerous churches were built. Monasteries were also constructed, among which was a second one in the capital Chang'an. In 745 a Chinese decree mentions the existence of a new monastery in the other imperial capital, Luoyang. There were yet other monasteries in more distant provinces, for example, at Chou-chih, 240 km north-west of Xi'an. To the west there were also churches and monasteries in the town of Dunhuang, an important stop on the Silk Road, and also in some towns in the province of Ningxia (today an autonomous region of China) on the principal route leading from the west of the empire to Chang'an. In the south, small Christian communities existed in the international port of Canton, and probably also in that of Quanzhou (the ancient town of Zaitun, to the north-east of Canton). In central China it is not likely that there were many Christian communities; at the moment it is only at Chengdu (to the south of the Yellow River) that a Christian presence can be confirmed under the Tang dynasty.

Under the emperor Xuanzong (712–56) and at the end of the eighth century, the Christians enjoyed several decades of prosperity. It was in this period that several bishops of the Church of the East were sent to China. They travelled by the land route along the Silk Road or by sea via Canton. It was the patriarch Sliba-Zkha (714–28) who raised the diocese of China to metropolitan status (the province of Beth Sinaye in Syriac, or Sinistan in Persian). Under the patriarch Timothy I (780–823), Thomas of Marga mentions the name of a metropolitan enthronized for China in this period, David, a monk of the monastery of Beth Abhe (situated 80 km north-east of Nineveh/Mosul).

Decline from 845

From 635 to 845 Christianity in China enjoyed a period of stability. But the rate of its development could not rival that of Buddhism, which expanded very rapidly. The two emperors Daizong (762–80) and Dezong (780–805) continued to support the Christians of the Church of the

East. But although the imperial power up to that time approved of the new Christian religion, that situation changed in 845. In fact, between 843 and 845 xenophobic reactions orchestrated by Chinese Taoists resulted in the emperor Wuzong (840–6) publishing an edict in 845 that outlawed all foreign religions in China and ordered the closure of their monasteries. The Buddhists were the first to be affected by these measures, but the Manichaeans and the Christians suffered equally from the interdicts. The emperor Wuzong ordered the expulsion of the Zoroastrians and the Christians from China, which may explain why the Xi'ian Stele was buried at that time.

This persecution of foreign religions was very severe and accounts for the decline of the Church of the East in China. In 877 there is mention of massacres of thousands of Christians at Guangzhou (Canton, since antiquity the biggest town in southern China, today 1213 km to the southwest of Shanghai). Christians were also killed elsewhere, for example, at Quanzhou. It is thus be understandable why the name of the Chinese metropolitanate of Beth Sinaye was omitted from the list of metropolitan provinces compiled by Elias of Damascus in 893.

In point of fact, after 845 very little is heard of Christianity in China, even though there is some indication of its survival for a while, for example, at Guangzhou around 878–9, or even after the fall of the Tang dynasty in 907. After the ninth century it appears that the Church of the East survived only in the north-western part of China, most of all among the Keraite, Öngüt and Uyghur tribes. There was still perhaps a Christian monastery at Xi'an in 1076. The reports of Arab travellers give some indication of Christians, churches and even monasteries in the tenth and eleventh centuries.

With the official disappearance of East Syriac Christianity that followed the expulsion of its members by the Tang in the ninth century, the question may be asked as to what happened to those Christians who would still have remained. We know that the Church of the East had no official presence in China under the Song dynasty (960–1272). For the intermediate period between the Tang and Yuan dynasties (mid-ninth to the end of the thirteenth century), academic research is continuing on the basis of historical archives and archaeological discoveries to try to establish the existence and location of Christian communities, for example, in the Tangut kingdom of western Xia (1038–1272) in north-west China and Mongolia. According to Hidemi Takahashi, during the twelfth century there were Christians of the Syriac tradition in China, thanks to movements in the interior of northern China of Turco-Mongolian tribes converted to Christianity. According to Matteo Nicolini, we know of Christians in Canton at the end of the eleventh century who were perhaps merchants originating from Persia. Karel Pieters notes that the fact that Christian gravestones have been discovered dating from the Song (960–1279) dynasty and the Liao (907–1125) dynasty, also known as the Khitan Empire, reveals that there were some Christians remaining in China in the north, in Inner Mongolia. The Christian remains that have been found from these periods ought to be attributed principally to the Öngüts. The tombstones that have been discovered bear inscriptions in the languages and characters of Syriac, Uyghur and Chinese.

Under the Mongols in China (1260–1368)

At the end of the thirteenth century the Mongols seized power in China. The great Mongol khan Kublai (1260–94) was the founder of the Yuan dynasty (1271–1368). In 1267 he fixed his capital at Khanbaliq, 'the khan's town' (or Dadu, the 'great capital'), which is the modern capital of Beijing. The Church of the East benefited from the Mongol conquest. It was able to return to China and be active again there. Some churches and monasteries were constructed anew in the large towns and the ports until the end of Mongol rule in China. It is under the Yuan dynasty that the presence of the Church of the East in China reached its peak, in particular thanks to the Christian Turco-Mongol tribes such as those of the Keraites and the Öngüts. People from central Asia, among them some Syriac Christians of Persia and Central Asia, also established themselves in China.

At the end of the thirteenth century two new metropolitan provinces were created in China by the hierarchy of the Church of the East: Tangut (in the north-west provinces of modern China, with its capital in the modern town of Yinchuan), and also Katai-and-Öng (in the north of China in the region of the Christian Öngüts around the great curve of the Yellow River). The metropolitans Giwargis and Nestorius of Katai-and-Öng are mentioned in the biography of the patriarch Yahballaha II (1190–1222). The name 'Katai' (Cathay) is derived from the tribe of the Qara-Khitai, whose kingdom was abandoned to the Mongols after the latter killed their leader Küchülüg in 1218.

The location of the suffragan dioceses in this period is not known with any precision. One of them (under the metropolitan of Katai-and-Öng) was very probably situated at the Öngüt capital at Tung-sheng/Dongsheng, and others perhaps in the large towns of Liangzhou and Su-chou (province of Gansu) in the metropolitanate of Tangut. In northern China in 1253 the existence is known of a bishop at Segin (the modern town of Ta'tung/Datong in the province of Shanxi), at that time known under the name of Hsi-Ching/Xijing, 'the western capital'. In southern China the bishops of the Church of the East resided in the ports of Hangchow/Hangzhou and Ch'uan-Zhou/Quanzhou, where the presence is attested of a bishop called Shlemun of Manzi (in southern China), who died in 1313, as is indicated on his gravestone, which has been discovered.

We do not know much about the number of people converted to Christianity in this period, but it may be asserted that their number was large enough to have their own metropolitan sent to them. Among the Christians of the Church of the East who returned to China at that time there were monks, merchants and administrators. Some Christians of the Church of the East occupied important positions in the Yuan Mongol government, for example, that of local governor (*darugha*). Thus we know of the existence of a Christian originally from Samarkand called Sargis (Sergius) who was appointed governor of Chinkiang/Zhenjiang (in the province of Jiangsu, a port on a Yangtze River) in 1278. A little later Sargis built six monasteries at Chinkiang and another at Hangzhou. He was also a priest and is mentioned in the writings of Marco Polo. In 1295 another monastery was built in the same town by a Christian called Mark. According to David Wilmshurst, during the Yuan period many churches and monasteries were built in China but there were relatively few Christians. Marco Polo lived in China from 1271 to 1293. He mentions Christian communities of the Church of the East in most of the towns that he visited, often consisting merely of a few families. Similar small colonies of Christians were found at the time in most of the principal towns. A special department was established by the government to supervise their activities. It is mentioned for the last time in 1351.

Some local Chinese histories mention the existence of Christian churches and monasteries in their respective regions. The Church of the East was able to flourish in the Empire during the first half of the fourteenth century. After the mid-fourteenth century there are few references to Christians in China of the Yuan. In fact in 1368, when the Chinese Ming emperors took the capital, Khanbaliq, from the Mongols, all the Christians were expelled at the same time as their Mongol protectors. Subsequently, China remained isolated from the rest of the world for a long period. It was therefore with the fall of the Mongols in China that Christianity disappeared officially in this country for the second time. The two metropolitan provinces of the Church of the East in China (Tangut and Katai-and-Öng) ceased to exist. Some scholars think, however, that in northern China, among the Keraites and the Öngüts, small communities of Christians possibly survived until the fifteenth century. But there is no solid evidence after the fall of the Yuan dynasty in the middle of the fourteenth century. In 1582 the Catholic Jesuit missionaries who arrived in China from Europe found only some rare traces of the memory of a Christianity of the Church of the East surviving in China.

Archaeology and Literary Sources

The presence of Christians of the Church of the East in China is confirmed in an undeniable manner by some written documents, not only in Syriac and Chinese but also in other languages, including Arabic and Latin. These literary sources are limited today to a small number of texts. That is why an attempt to reconstruct the history of the Church of the East needs to take in as many other sources as possible, including archaeological discoveries. Within this perspective, inscriptions have been found on steles and tombstones, and also crosses, various artefacts and some mural paintings. Christian inscriptions have been found in various languages, for example, in Syriac, Mongolian, Chinese, Sogdian, Old Turkic and Middle Persian, and written in different scripts (Syriac and others).

Apart from the monumental Xi'an Stele already mentioned, other ancient steles have been discovered in China. Numerous gravestones have also been recovered in Inner Mongolia, in Xinjiang and in the towns of the south-east, such as Yangzhou/Yang-chou, where a gravestone (unearthed in 1981) dedicated to the Lady Yelishiba (Elizabeth), who died in 1317, was written in Syro-Turkic and Chinese. A large number of gravestones have also been found at Chuan-chou/Quanzhou, a prefecture town of the province of Fujian, and in other towns with inscriptions in Syriac and Chinese. On the steles and some gravestones decorative elements of a Chinese type may be seen, such as dragons and lotus flowers. All the gravestones and other objects that have been unearthed shed a light on the geographical expansion of Syriac Christians in China.

When the scattered mentions of the presence of Christianity in Yuan China are brought together, a concentration of Syriac Christians may be identified in three principal sectors: the north-west region (the modern provinces of Xinjiang, Gansu, Inner Mongolia and a part of Shanxi); the region of Jiangnan (eastern China, south of the Yangtze River), in the port cities; and also around the Yuan capital of Khanbaliq (modern Beijing). Here I shall cite some of the principal Christian archaeological discoveries made in China, some of them in places already mentioned in Chapter 4 on the Silk Road but situated today in China.

In the modern province of Xinjiang (a region in north-west China) important remains of churches, monasteries, books and other objects connected with Christian life that have been found in the oasis-towns testify to the extent and vitality of the Church of the East in this region.

At Almaliq (Xinjiang, <https://en.wikipedia.org/wiki/Almaliq>, Xinjiang) there were 'Nestorian' Turks living there in the Mongol period. The town was an important centre of the Church of the East and the seat of a metropolitan. A very great number of gravestones have been discovered with inscriptions in Syriac and Turkic, most of them dating from 1250 to 1345, which demonstrates the existence of a large Christian community in this period. Eight steles have also been discovered there.

At Ürümqi (today the capital of the autonomous Uyghur region of Xinjiang, 193 km north-west of Turfan) bas-reliefs with crosses among other objects have been discovered.

In the Turfan Oasis, on the northern edge of the Taklaman Desert, an important stop on the ancient Silk Road, there are several archaeological sites of importance for the history of the Church of the East. The most striking discovery has been made among the ruins of an ancient Christian monastery at Bulayiq (10 km north of the modern town of Turpan/Turfan). It was there that numerous manuscripts and manuscript fragments were discovered in 1905 by a German expedition led by Albert von Le Coq. Fragments of manuscripts written in several languages were found, including Sogdian, Syriac, New Persian and Old Turkic which give enormous insight into the liturgical and ecclesiastical traditions of the Church of the East. This collection of texts may be considered a wonderful Christian library dating from the ninth to the thirteenth/fourteenth centuries. It includes biblical and liturgical texts, hymns, psalters and also historical and ascetical texts. These writings testify to the links that were maintained with the mother-Church of Mesopotamia, whose patriarchal seat was at Baghdad. In our own day these manuscripts are preserved for the most part at the Berlin-Brandenburg Academy of Sciences and Humanities (Berlin-Brandenburgische Akademie der Wissenschaften), at the Berlin State Library (Staatsbibliothek zu Berlin), and also at the Asian Art Museum (Museum für Asiatische Kunst), all in Berlin. A scientific project is in the course of cataloguing these manuscripts fragments, of which the Christian fragments number about 1100. Nicholas Sims-Williams, Erica Hunter and Peter Zieme have catalogued the manuscripts and there have been many publications on them.

Also in the Turfan Oasis (30 km south-east of the modern town of Turpan) lie the ruins of Gaochang (the ancient town of Qocho). It was here in 1905 that Albert von Le Coq discovered a church of the East Syriac tradition dating from the seventh/eighth century and decorated with mural paintings.

At Hami (410 km to the east of Turfan, in eastern Xinjiang) Christian objects that have been found (dating from the ninth to the fourteenth century) show that Christianity was well established in this region, in the Uyghur kingdom of Qocho (<https://www.cambridge.org/core/books/history-of-inner-asia/uighur-kingdom-of-qocho/56C803A6DC3EF2D682A6AE921169B872>).

Seven hundred and ninety-nine kilometres to the south-east of Turfan and 14 km to the south-west of the modern town of Dunhuang, another famous town on the ancient Silk Road (today in the province of Gansu in north-west China), to the east of the Taklaman Desert and to the south-west of the Gobi Desert, are the Mogao Caves (also called the Dunhuang Caves), which form a system of nearly five hundred Buddhist temples (hence today their alternative name: the Grottos or Caves of a Thousand Buddhas). In these caves thousands of manuscripts were discovered, some of them Christian, written in Chinese, Sogdian and Syriac. Aurel Stein bought some in 1907 and Paul Pelliot in 1908. Also found were paintings and objects that could date perhaps from the fourth to the eleventh centuries. These for the most part are of the Buddhist, Confucian and Taoist traditions that predominated in ancient China. A painting on silk (end of the ninth century) representing a man with a cross on his crown (perhaps Christ or a saint?) is kept at present at the British Museum in London. The presence of Christians is attested here until the fourteenth century.

To the south-west of the Gobi Desert, in the historical province of Gansu, the ancient town of Qara Qoto (whose ruins lie on the east bank of the River Ruoshui in the Badain Jaran Desert, on the southern edge of the Gobi Desert, on the Alxa plateau, in Inner Mongolia) was discovered in 1908–9 by the Russian explorer, Piotr Kozlov. Texts were also found there written in Syriac, Uyghur, Chinese and even Tangut (the language of a Sino-Tibetan group, according to Mark Dickens). This shows that there existed there a community of the Church of the East at the frontier of Mongolia to the north and at the frontier of the kingdom of the Tanguts (1038–1227) to the south-east. Fragments of manuscripts were found there – mostly liturgical texts of the East Syriac tradition in the Turkic language written in the Syriac *estrangelo* script (the most ancient version of the Syriac alphabet) dating from the thirteenth century.

Further east, in Inner Mongolia (today in north-east China) were the Öngüt people. Here numerous gravestones with inscriptions have been unearthed at the sites of Olon-Süme, Bailingmiao and Wangmuliang. At Chifeng in the east of Inner Mongolia a very beautiful tombstone has been found bearing an inscription in Syriac. At Olon-Süme (to the north-west of Hohhot, the modern capital of Inner Mongolia), which was the northern capital of the Öngüts in the thirteenth and fourteenth centuries during the Yuan dynasty, Christian ruins and cemeteries have been discovered (with inscriptions on the tombs) as well as a church of the East Syriac tradition. Recently a team from the Institute of Archaeology of Inner Mongolia has made some important discoveries of manuscripts.

Near Wuchun (about 70 km south-west of Beijing) stands the Daqin pagoda. The building has been identified by some as a monastery of the Church of the East perhaps founded in about 650. It is a tower of seven storeys. The monastery was reconstructed as a Buddhist monastery in 963. At present it is a Taoist shrine.

In the village of Fangshan (about 50 km to the south-west of Beijing) two blocks of stone and two steles have been found with crosses and inscriptions in Syriac. As indicated by an inscription citing an imperial edict, this place was the 'monastery of the Cross'. The site was later transformed into a Buddhist temple. Today some people still call it the temple of the Cross.

To the east of Xi'an, at Luoyang, the ancient eastern capital of the Tang dynasty, funerary crosses have been found and above all a stele, discovered in 2005 (and dated 814/15), which appears to be as important as that of Xi'an, with inscriptions that inform us about a local monastic community.

In the south-west of China, at Quanzhou (or Zaytun, the medieval name of Quanzhou), near the South China Sea, several Christian gravestones have been discovered of the tradition of the Church of the East, dating from the thirteenth and fourteenth centuries, with crosses. There and in the environs of this commercial port that was very active in the fourteenth century, Christian archaeological material has been discovered which is at present exhibited in Quanzhou's Maritime Museum.

Christian archaeological finds may be viewed in several other museums in China, including Beijing. In Hong Kong, the university museum (the Fung Ping Shan Museum) possesses a collection of hundreds of bronze crosses of the fourteenth century (from the Yuan period and the Öngüt region) which was assembled by a British in the north of China in the 1920s.

Concluding Remarks

At the present time we need to rediscover the fascinating history of the Church of the East in China and its missionary activities. Since 1979, when China opened its doors, the interest of some Chinese in this heritage has been striking, notably of experts who have been attracted to the history of ancient Christianity in their own country. This research is at present conducted not only by Chinese scholars who carry out archaeological excavations and study the inscriptions, but also by foreigners. Conferences have been organized on these subjects followed by publications. We may note in particular the conferences organized regularly since 2003 on the Syriac Christianity of China and Central Asia by the University of Salzburg (see their publications in the bibliography). In 2015 a conference on the Syriac studies at Turfan and dedicated to Sogdian culture, entitled 'From Tajikistan to Turfan: Traces of the Cultural Heritage of the Sogdians', was held at the National Museum of Tajikistan at Dushanbe (<https://www.soas.ac.uk/news/newsitem105798.html>). In the same year a conference was organized in Hong Kong by Hong Kong University, entitled 'The International Conference on Jingjiao' (<https://www.hkihss.hku.hk/events/jingjiao2015/program.html>).

Since the end of the Cultural Revolution in 1976, China has experienced an unprecedented number of conversions to Christianity. In our own day most of the Christians in China are Protestants (belonging chiefly to the Evangelical movement). There is also a large number of Catholics. The ancient history of the Church of the East in their country could interest all these Chinese who wish to study their Christian roots.

The first representative of the Church of the East to return to China in 1996 was Mar Giwargis Sliwa, the present patriarch of the Church of the East, at that time metropolitan of Baghdad. He visited Shanghai and Beijing, as well as the temple (or monastery) of the Cross at Fangshan. He also visited the pagoda of Daqin and the Steles Museum at Xi'an, which issued a communiqué to the press. In 2010 Metropolitan Awa, bishop in California and secretary of the Holy Synod of the Church of the East, went to Hong Kong to take part in a conference whose theme was the 'Luminous Religion' (Jingjiao). A liturgy was celebrated on that occasion at the Lutheran theological seminary – the first on Chinese territory after very many centuries.

All these events open doors for the development now of new relations between the Chinese and the Church of the East. This is what is happening in the case of the Jingjiao Fellowship at Hong Kong. Its director, David Tam, has translated into Chinese the most frequently used text of the Liturgy of the Church of the East, the Anaphora of Addai and Mari.

Today both Chinese and foreigners are discovering the pleasures of tourism in China. There are numerous visitors to the famous sites mentioned above, for example, the Steles Museum in Xi'an and the sites of Turfan and Dunhuang.

To conclude, neither the written sources that have survived nor the archaeological discoveries that have been made give us sufficient information to allow us to form a complete picture of the state of East Syriac Christianity in China. Contemporary scientific research on the topic of the Church of the East is still at an early stage. Much remains to be done.

CHAPTER 6

Under the Mongols (1206–1368) and Tamerlane (1370–1405)

Under the Mongols (1206–1368), the Church of the East enjoyed a last period of expansion, thanks to the Mongols' unification of Asia from the Euphrates to the Yellow Sea. During this period the Church of the East planted itself again in China. It was at this time that the Church of the East attained its broadest geographical extent.

The Mongols

In the thirteenth and fourteenth centuries the Mongols came to establish an Asian empire that was transcontinental and even European, one of the most extensive empires in history. In 1258, with their capture of Baghdad, they were for a time the masters of Asia. Beyond the Mongol Empire the principal states in the Middle East at that time were the Byzantine Empire (330–1453) and the Mamluk state in Egypt and Syria (1250–1517).

Even though the Mongols conducted ceaseless warfare, and despite their savage destruction of towns and populations, they also promoted a certain kind of 'universal' peace, the so-called Pax Mongolica, over their very vast territories in practically the whole of Asia. This period of stability created favourable conditions for travellers along the Silk Road and permitted the migration of peoples across Central Asia and China. The Mongol conquest resulted in an unprecedented unification of Eurasia that facilitated economic and cultural exchange. Thus from the social, cultural and economic viewpoints, the unified administration of this empire had a stabilizing effect in Asia and beyond. Thanks to this order that was imposed in Asia in the second half of the thirteenth century by the Mongols, the Church of the East under their protection was able to grow and prosper.

The Geography and Structure of the Mongol Empire

The names of the first Mongol leaders and their successors who played a role in the history of the Church of the East are as follows.

Temüjin, the future Genghis Khan (d. August 1227), succeeded in unifying and bringing under his control not only the Mongols but also nomad tribes of north-east Asia such as the Keraites (to the south of Lake Baikal, where the great Mongol Khan was to fix his capital Karakorum) as well as their neighbours the Naimans (to the north of the Altai mountains), the Merkits, the Öngüts and the Uyghurs (in the Turfan Oasis and the Tarim valley, to the south of the Altai). After he was proclaimed their chief, Temüjin received the title Genghis Khan, which means 'universal sovereign'. It was he who established the Mongol Empire.

Genghis Khan's vast empire was divided into several kingdoms or domains (*ulus*), territories which in the west were called khanates, and which were assigned to his sons and successors. Their chiefs (khans) were supposed to obey the great khan (*khagan*) or emperor. The eldest son of Genghis Khan, Jochi (d. February 1227), withdrew to the Ural and Turgai steppes which had been assigned to him by his father. His domain was shared out among his own sons Batu and Orda. It was Ögedei (1229–41), the third son of Genghis Khan who was designated great khan. He controlled the territories east of Lake Balkash as far as Mongolia. In 1235 he established his capital at Karakorum, where archaeologists have found a building which could be a church dating from this period. The ruins of this ancient town are in Mongolia near the modern town of Kharkhorin (in the north-east of the province of Övörkhangai in the centre of Mongolia) and not far from the Buddhist monastery of Erdene Zuu, which for centuries has been the most important Buddhist shrine in Mongolia.

Jagatai (d. 1242), Genghis Khan's second son, was given Central Asia and northern Iran, the so-called khanate of Jagatai, in Transoxiana, between the rivers Amu-Darya and Syr-Darya (modern Uzbekistan), and the region around Kashgar. He established his capital at Almaliq (today in the autonomous region of Xinjiang, in north-west China). Tolui (1192–1232), the youngest son, inherited the northern Mongol territories (modern Mongolia) the original paternal patrimony. Möngke (1251–9), the eldest son of Tolui, was proclaimed the fourth great khan. Tolui's fourth son, who succeeded his brother Möngke as great khan of the Mongols, was Kublai Khan (1260–94). It was he who was the founder of the Yuan dynasty (1271–1368), which reigned over the territories lying today in China, Mongolia and regions in between. Kublai became the first non-Chinese to conquer China, making his capital Khanbaliq (the modern capital, Beijing).

In the south-west of the empire, Hulagu Khan (1256–65), Möngke Khan's younger brother, accomplished the conquest of Persia, occupied Baghdad (1258) and abolished the Abbasid caliphate. Hulagu was named the head of the ilkhanate (that is to say, the 'vassal khanate') whose territories were principally in Persia and the surrounding regions. He fixed his capital at Maragha, then at Tabriz and Soltaniyeh (lying between Tabriz and Tehran). This is the so-called Ilkhanid dynasty.

In the north-west, Jochi's descendants conquered the Russian steppes to the west of the Irtysh River (the name meaning 'white river'), which rises in the Altai mountains and then runs through western China before crossing Kazakhstan and Siberia. The khanate of Orda is known historically as the White Horde. The khanate of Batu is known under the names of the khanate of Kipchack and of the Golden Horde (1243–1502). Batu's camp was in the territory of the lower Volga (to the north of Astrakhan), a territory lying between Lake Balkash and the Black Sea, and exercised direct control over the region to the west of the Urals. The Mongol armies invaded Rus' (1236–40), which included, among other areas, some territories of the modern countries of Russia, Ukraine and Bielorussia. This was followed by a brief invasion of central Europe (1240–2). On the news of the death of the great khan Ögedei, Batu and his troops turned back to the east, thus liberating Europe. Batu returned to his encampment and died in 1255 in his capital Sarai (established in 1243 on the lower Volga, 120 km north of Astrakham, in modern Russia). The khans of the Golden Horde ruled over western Siberia and the south of what was to be Russia until the end of the fifteenth century (1243–1502). The khan Berke (1257–67) was the first Muslim Mongol sovereign of the Golden Horde.

On the death of Genghis Khan in 1227 the Mongol Empire, in geographical terms, extended from the Yellow Sea to the Caspian Sea. The greatest expansion of the Empire came in the time of Ögedei Khan (1229–41), taking in the modern territories of Mongolia, Afghanistan, Iran, Iraq, Azerbaijan, Armenia, Georgia and Ukraine, as well as large parts of Siberia, southern Russia and Turkey, together with the regions of the north and west of China as far as Tibet. We should note that in 1247 Georgia was divided into two kingdoms by Güyük (1246–8), the third great khan and eldest son of Ögedei Khan. With regard to the Armenian kingdom of Cilicia (today in south-east Turkey), its king, Hethum I (1226–70), had his status as a vassal confirmed in 1254 by Möngke Khan at Karakorum.

Religions

The original religion of the Mongols was shamanism. In the thirteenth century, in the empire of Genghis Khan, several different religions coexisted, principally Buddhism, Christianity, Manichaeism and Islam. A number of the Mongol khans were Buddhists, and later Muslims. During the first period of Mongol rule there was religious tolerance and a harmonious cohabitation of all the great religions. In the period that followed that of Hulagu, the ilkhans increasingly adopted the Tibetan form of Buddhism, while at the same time tolerating different religions, including shamanism and Christianity.

With regard to ecclesiastical matters, at the beginning of the Mongol era the patriarchs of the Church of the East took advantage of these political circumstances that seemed to them favourable at that time for reorganizing and strengthening their Church in Central Asia and China. In fact, under Mongol rule the Christians enjoyed a privileged position. In what way was that? As we have already seen, at the beginning of the thirteenth century Christianity was flourishing thanks to the expansion of the Church of the East in Central Asia and in particular in the Mongolian steppes where a large number of Turco-Mongol tribes (Keraite, Naiman, Oirat, Merkit and Öngüt) became Christian, including their ruling families.

Contacts between Mongol Rulers and the Christians of the Church of the East

In the Mongol Empire the Christian religion received an official organization and the Christians of the Church of the East were able to enjoy a number of privileges. Under Genghis Khan and his first successors, many of them came to be employed by the Mongol rulers and exercised a certain degree of influence at the heart of the empire. Among them were merchants. Others became doctors at the court or high functionaries, governors and even generals in the army. All these Christians contributed without doubt to the diffusion of Christianity in the vast territories administered by the Mongols. In Persia and China, the two principal poles of the Mongol Empire, the towns enjoyed a considerable period of growth due above all to international commerce. The various contacts between the Mongol rulers and the Christians of the Church of the East had certain consequences, as the following example will show.

First of all, several Mongol sovereigns were sons of Christian mothers, sovereigns such as Möngke, Kublai and Hulagu. In fact for several generations a number of Mongols married Christian wives. Genghis Khan's family practised mixed marriages with the royal families and the élite of the Turco-Mongol tribes who adhered to the Christian religion in order to conclude alliances. Thus Christianity penetrated Genghis Khan's family. After himself marrying a Keraite Christian, Genghis Khan gave his daughter Alaqai Beki in marriage to the son of Alakush, an Öngüt khan who was probably Christian. The family bonds between Genghis Khan and the Öngüts remained strong, as is shown by the case of King Körguz (d. 1298), also known under the name 'George'.

It is especially from the fact that they had Christian wives that the Mongols were well disposed towards the Christians of the Church of the East. And it is principally through the mediation of these Christian women, princesses and queens, parents or spouses of the Mongol khans, that the Church of the East made an impact in the Mongol Empire. The woman who was the most influential and most celebrated in the history of the Mongol Empire was Sorgaqtani (d. 1252), the wife of Tolui. She was a Christian Keraite princess and did her best to support her Church. Sorgaqtani was the mother of two great khans, Möngke and Kublai, and also of Hulagu, the founder of the Persian ilkhanate. She had a great deal of influence over her sons, who respected her Christian faith and those who shared it. The wife of the khan Hulagu, Doquz (d. 1265) was also a Christian Keraite princess, and a cousin of Sorgaqtani. She too exercised a strong influence on her husband and obtained permission from him for the construction of several churches. Throughout her life she showed concern to protect the Christians, who as a result enjoyed a situation in which they were able to flourish. In 1258, after Hulagu had conquered Baghdad, the fact that the Christians were spared was probably due to her intercession on their behalf. One of the caliph's palaces was at that time offered to the patriarch of the Church of the East as his residence. In fact, when the Mongols took Baghdad they promised the patriarch of the Church of the East, Makkikha II (1257–65) to spare the Christians not only in Baghdad but also in Tabriz, Aleppo, Damascus and elsewhere. It was also to the support of Doquz that Denha I (1265–81) owed his nomination as patriarch. The wife of Möngke was also a Christian who belonged to the Oirat tribe. Möngke allowed Christian ceremonies to be celebrated at court at his wife's request. Töregene, a Christian Naiman princess, was given in marriage to Ögedei. It was this queen (*khatoun*), Törgene, who assumed the regency after her husband's death until the election of their son Güyük as the third great khan in 1246.

Some Mongol khans occasionally became Christians and were baptized; others even baptized their children. This was the case with the great khan Güyük, who when he came to power chose a Christian minister. The great khan Möngke was without doubt also baptized. The son of the Christian princess, Sorgaqtani, he always treated the Christians with benevolence, probably out of filial respect and love. With regard to the third ilkhan of Persia, Teküder (1282–4), he was baptized in his infancy, but as an adult was the first ilkhan to be converted to Islam, taking the Muslim name Ahmed. Oljeitu (1304–16), son of Arghun and Uruk, a Christian 'Nestorian' princess, was baptized under the name of Nicholas by the patriarch of the Church of the East. As for the ilkhan Hulagu, he was due to be baptized but finally converted to Buddhism. Following his conquest of Persia, he favoured the Christians in his new kingdom. We must record above all the name of Sartaq (1256–7), who was baptized and remained a Christian. He was the son of the khan Batu and succeeded him as the khan of the Golden Horde, but his reign was very short. With regard to Güyük, the third great khan (1246–8), the historian Bar Hebraeus (died in 1286 at Maragha), who was the maphrian of the Syrian Orthodox Church, writes that he was a Christian. Güyük was sympathetic to the Christians. Apart from the khans, some notables were Christians, such as prince Nayan (d. 1287), related to the family of Genghis Khan and a distant cousin of Kublai Khan.

At Court and in the Administration: Doctors, Translators, Governors, Army Generals and Clerics

As a certain number of great khans and Mongol princes were raised by Christian mothers and Christian teachers of the Church of the East, this was no doubt a strong incentive for them to choose Christians as scribes, doctors and others. These connections of all kinds that the khans maintained with the Christians in their entourage enabled them to avoid any fanaticism and any religious crisis in their multi-confessional empire. At the court and in local civil administration the Mongols promoted Christians to various positions of trust. Thus the great khan Möngke had a Christian as his chief secretary (the equivalent of a chancellor or minister), the Keraite Bolgai. Under the Yuan dynasty we have already mentioned the governor of the important town of Zhenjiang (near the Yangtze River), Mar Sargis, who was also a priest. Under the same dynasty, a very important personage at the court was a Christian doctor, a member of the Church of the East born in Syria named Ai Xieh (1227–1308), the Chinese equivalent of Isha, or Jesus. Kublai Khan appointed him head of his department of astronomy and medicine. He served, moreover, as interpreter for a high-ranking delegation to the ilkhan Argun at Baghdad. In 1291 he was appointed head of the office for the Christian religion and in 1297 became the equivalent of a minister. His five sons also attained high office.

Christians had influential posts not only at court but also in the army, even among the military leaders. For example, Chinkai (an Uyghur) was first secretary–minister, under Ögedei and then under Güyük. And Qadq (a Naiman) was first secretary–minister under Ögedei. He was also Güyük's tutor and was a judge during his reign. In 1257 the campaign conducted against the Abbasid caliphate was directed personally by Hulagu and his Christian general, Kitbuqa, of the Naiman tribe.

With regard to the clergy, we have sure knowledge of examples of good relations between the Mongol leaders and the clergy of the Church of the East – which included without doubt a certain number of Mongols. The great khans and the khans had priests of this Church at their court and in their entourage. For example, we know that the tutor of Möngke's son was a priest of the Church of the East called David and that the priests celebrated the Syriac liturgy in a tent used as a church close to that of Möngke.

The History of Mar Yahballaha III and Rabban Sauma

The two most celebrated figures of the Church of the East under the Mongols, who enjoyed a very great reputation, were two monks, Sauma and Markos (Mark), the latter of whom became the patriarch Yahballaha III (1281–1317), a name which means 'Given by God', the Syriac equivalent of Theodosius. Both of them probably belonged to the Öngüt tribe. In around 1275 the two monks decided to make a pilgrimage to Jerusalem, a holy Christian place which they never managed to reach on account of the insecurity caused by the war in Syria and Palestine between the ilkhanate and the Mamluks. The story of their fascinating travels may be read in an account written in about 1317. This chronicle, composed by the monk Sauma, has been published in English under the title *The Monks of Kublai Khan, Emperor of China*. It is worth summarizing here this unusual account.

After leaving Khanbaliq (modern Beijing), the two monks headed towards the region of the Tanguts. They made their way to Khotan and Kashgar (today in the autonomous region of Xinjian, in north-west China) and then to Talas (today Taraz, in Kazakhstan). In all these places the presence of Christians is documented by other travellers of the era such as Marco Polo in his *Book of the Wonders of the World* (in Italian, *Il Milione*) composed in 1298, which describes not only his travels across Asia in 1271 and 1295 but also his experiences at the court of Kublai Khan at Khanbaliq.

In 1280 Sauma and Markos arrived at Maragha, and later at Baghdad. They met the ilkhan Argun, the Mongol ruler in Persia, and also their patriarch Denha I (1265–81). In the same year the patriarch Denha enthroned Markos under the name of Yahballaha as metropolitan of Katai-and-Öng, in Öngüt territory in the north of China. But Yahballaha was unable to return at that time because of armed conflicts barring the route. In 1281, on the death of Patriarch Denha, Yahballaha was elected patriarch of the Church of the East. No doubt this choice was made for political reasons. In fact, the patriarch's origin allowed for certain advantages to be expected through his access to the Mongol suzerains, whose language and customs he was familiar with. The patriarch Yahballaha was thus in a position to serve his Church thanks, too, to his contacts with the court of the ilkhan Mongols. At the enthronization of the patriarch Yahballah III in Baghdad in 1281, the presence is notable of twenty or so metropolitans, among them the metropolitan of Tangut (China). In thirty-six years this patriarch ordained and enthroned seventy-five bishops and metropolitans.

With regard to Sauma, this polyglot became a bishop in 1287–8 and was sent as an ambassador by the khan Argun and his patriarch to Europe to form an alliance with the principal Christian rulers, principally against the Mamluks (1250–1517), in order specifically to conquer the territories of Syria and Palestine and recover Jerusalem from the Muslims. In sending a Christian envoy, it was thought that this initiative would have more chances of success, but the project bore no concrete results. Thus it has been established that in the Mongol era contacts between Central Asia and Western Europe were intensified. In 1291 the Muslims captured the town of Acre, the last fortress in the region held by the Crusaders.

To record some interesting facts, on his journey to Europe Sauma stopped at Constantinople, Rome, Genoa and Paris. Through Sauma the Church of the East enjoyed a significant diplomatic and political role because in the course of his journey he met the emperor Andronikos II Palaiologos (1282–1328) at Constantinople, King Philip IV of France, Edward I, king of England, and also important ecclesiastical figures such as Pope Nicholas IV. In Rome Sauma was in a position to explain that Christianity had been brought to the East by the apostles Thomas, Addai and Mari and not by the Roman Catholic Church, that the missionaries of the Church of the East had evangelized the Mongols, the Turks and the Chinese, and that some of the queens and children of the Mongol kings had been baptized and had churches (for the most part, in fact, tents with portable altars) in their camps. Sauma died in Baghdad in 1294.

With regard to Patriarch Yahballaha, he left Baghdad and moved to the north to Maragha (639 km away), the new capital of the Mongol ilkhanate. He and his community experienced various trials. In fact they were victims of frequent acts of harassment, torture, massacre and destruction perpetrated by certain local Muslim leaders, against which Patriarch Yahballaha tried to defend his community and himself in vain. The chronicle mentioned above, *The Monks of Kublai Khan, Emperor of China*, is of capital importance for the documentation of the history of the Church of the East in China, Turkestan and Mongolia in the thirteenth century. It contains a great deal of information on the ilkhans of Persia and their relations with the Christians, and also on the situation of the Church of the East under the Mongols. It also describes certain events that brought on the collapse of the Church of the East in China, Central Asia and Persia.

The Mongol Expansion in the Middle East Stopped by the Mamluks. Islam becomes the State Religion of the Persian Ilkhanate in 1295

In 1259 Hulagu attacked the Ayyubid principalities of Syria, took the towns of Edessa and Nusaybin (Nisibis) and arrived at Aleppo. The Christians of Syria and Iraq considered the Mongol invasion a liberation from Muslim oppression. After the Mongols had captured the city of Baghdad (1258) and conquered Syria (1259–60), the Muslim forces appeared to have lost, except for the Mamluks in Egypt. But in 1260 at Ain Jalut (in Galilee, today in the north of Israel) the Mongol army, led by Kitbuqa (mentioned above) was defeated by the Mamluks of Egypt. The military progress of the Mongols, which seemed invincible, came to an abrupt halt. This was the first big defeat for the Mongols. It marked the end of their western advance. In 1312 the Mongols suffered another bloody defeat at Rabat in Syria, which put a definitive end to the prospect of an alliance between the Mongols and the Christian powers against the Mamluks. All hope of establishing a Christian power in the Middle-Eastern region in coalition with the Catholic powers of the West and the Crusaders (1095–1291) was crushed. In 1322 a formal peace treaty was concluded between the ilkhanate and the Mamluk sultanate of Egypt.

In fact, given that the Mongol forces advanced increasingly westwards, it was inevitable that they would find themselves face to face with the Muslim powers that had established themselves in the Middle East, and so with Islam. The first three generations of Mongol conquerors had not adopted the religion of their new Muslim subjects. But when the Mongols of Persia (the ilkhans of the khanate of Jagatai), along with their cousins of the Golden Horde to the north, decided to become Muslims, the geopolitical and religious situation of these regions changed radically. It was in 1295 that Islam was adopted officially as the state religion of the ilkhanate by Ghazan (1295–1304), the seventh ilkhan of Persia, a great-grandson of Hulagu. The conversion of the ilkhans to Islam totally changed the policy of the Muslim ilkhans towards the Church of the East.

The Policy of the Mongol Ilkhans towards the Church of the East

Let us attempt now to understand the evolution of the policy of the Mongol ilkhans towards the Church of the East in more detail. In 1258, after the Mongols had abolished the Abbasid caliphate, there followed about three decades of relative security for the Christians of the Church of the East.

It was then that they were able to construct or restore several monasteries and churches, among them the monastery of St Augin near Nisibis and two churches at Maragha – that of St Challita and that of Sts Mari-and-George, the latter thanks to the support of the fifth ilkhan Gaikhatu (1291–5). In September 1295, however, an important event occurred: Islam became the state religion of the ilkhanate, which brought in its wake open persecution of the Christians. Finally, although the first decade of the fourteenth century was a period of peace, it was followed by a terrible massacre at Erbil (in northern Iraq, 368 km south-west of Maragha) on 1 July 1310.

What can one say about the principal ilkhans? After Hulagu captured Baghdad in 1258, sent by his father, the great khan Möngke, he designated the town of Maragha (today in north-west Iran, 638 km from Baghdad) as his new capital. The ilkhan Abaqa (1265–82), a Buddhist like his father, Hulagu Khan, was favourably disposed towards the Christian communities and married Maria, the daughter of the Byzantine emperor Michael VIII Palaiologos. Abaqa was the protector of the patriarch Denha I (1265–81) and his successor, the patriarch Yahballaha III. Argun (1284–91), a Buddhist, showed his favour to the Christians by offering them certain government positions, as his father had done and his grandfather Hulagu.

In 1294 the patriarch Yahballaha III laid the foundations of a great new monastery at Maragha dedicated to St John the Baptist, in order to make it the new patriarchal residence. His plan had to be abandoned in 1295 on the death of Gaikhatu (1291–5), Argun's brother, a Buddhist who was accused of having too much sympathy for the Christians. Baidu (who reigned in 1295), a cousin of Gaikhatu and grandson of Hulagu, showed himself at first favourable towards the Christians, but soon ended up adopting Islam, the religion of the large majority of his subjects. Although Baidu's son, Sartaq, was a Christian, his brother Berke, who became the Kipchak khan, was an Islamic sympathizer. As for Ghazan (1295–1304), the eldest son of Argun, he converted from Buddhism to Islam in 1295 and took the name 'Mahmud'. It was he who imposed Islam as the state religion. At the beginning of his reign Ghazan was influenced by a Muslim fanatic, the emir Nauruz. In 1295 at Maragha several churches were pillaged (among them the cathedral) or destroyed and the patriarchal residence was sacked. In the same year the patriarch Yahballaha III was arrested, tortured and imprisoned in his residence, which was repeated in 1296. Finally Nauruz was disgraced in 1297. In the second part of his reign Ghazan put an end to the persecution of the Christians, restored to them their traditional privileges and in 1303 even paid a visit to the patriarch at Maragha. The ilkhan Oljaytu (1304–16), Ghazan's brother, was baptized, but converted to Buddhism and then to Islam, upon which he pursued an anti-Christian policy. In 1304 he fixed his capital at Soltaniyeh. Although Oljaytu himself remained friendly towards the patriarch Yaballaha, Muslim hostility against the Christians increased in various regions of the ilkhanate.

Faced with the problems discussed above, the patriarch managed to flee and take refuge in the citadel of Erbil (in the north of Iraq, 368 km south-west of Maragha). Numerous Christians also withdrew to this stronghold which seemed their last possible refuge. But in July 1310 the capture of the city by the ilkhan's troops led to carnage despite the best efforts of the patriarch Yahballaha, which were of no avail. The four churches were destroyed, the patriarch's residence pillaged and numerous Christians massacred. Deeply shocked by this tragedy, the patriarch was forced to flee to Maragha, where he spent the last years of his life in the monastery, which became his residence. He died there in 1317 at the age of 73.

The last ilkhan who supported the Church of the East was Irinjin (in fact the Buddhist name of Gaikhatu, already mentioned) who was executed in 1319 by Abu Said, Oljaytu's son. After the death of the emir Chupan (d. 1327), a Mongol of Persia, the situation of Christians again became precarious and the monastery of St John the Baptist at Maragha was confiscated and transformed into a mosque. In Central Asia (Transoxiana and Turkestan) in the khanate of Jagatai, the Arab historian Ibu Fadhl Allah Al-Umari (d. 1349) says that the last wave of Islamization began in 1325. In this era numerous churches were destroyed in the east and many Christians became Muslims. After the reign of Abu Said (1316–35) the Mongol khanate in Persia came to an end. The Mongols were succeeded by Tamerlane.

Tamerlane (1370–1405)

Timur (1370–1405), better known by the name of Tamerlane ('the lame'), was a Turk who laid claim to Mongol ancestry. He was born in Kesh (in modern Uzbekistan, the ancient Transoxiana) some eighty kilometres south of Samarkand, in what was then the khanate of Jagatai. He established his capital at Samarkand. It is said that his military ambition was to restore the Mongol Empire. He ended up by conquering a great part of Central and Western Asia. He thus built up an immense empire based on military power and terror, with spectacular destructions and massacres. According to the historian René Grousset, he was a fanatical Muslim of extreme ferocity. His empire encompassed an immense territory extending from the Middle East to India and from the Persian Gulf to southern Russia, covering the modern countries of Uzbekistan, Kirghizstan, Tajikstan, Kazakhstan, Turkmenistan, Iran, Iraq, Syria, Azerbaijan, Armenia, Georgia, eastern Turkey, Afghanistan, Pakistan and as far as the sultanate of Delhi. Tamerlane even came close to the town of Kashgar, then in China. He was the founder of the Timurid dynasty, which lasted until 1507 – but his immense empire was torn apart by struggles among his numerous descendants and scarcely survived him.

Tamerlane's campaigns were immensely destructive: Christians were killed, churches and monasteries devastated, heavy taxes (*jizya*) and anti-Christian laws imposed. When his troops passed through Central Asia and beyond (in Azerbaijan, Iraq, Syria and Turkey), they wrecked the infrastructure of the Church of the East almost totally. Tamerlane conquered the khanate of Persia from 1381 to 1386. He took Baghdad in 1393, which the Ottomans in turn captured in 1533.

Concluding Remarks

On the whole, it may be said that the Christianity of the Church of the East played a relatively important role in the history of the Mongol empire, in particular through the mediatory role of various tribes principally of Christians of this Church. The thirteenth century was the last century in which the Christians of the Church of the East enjoyed a relatively calm existence, at least in the early years of Mongol rule. The Church of the East attained its apogee and its greatest geographical expansion in the era of the patriarch Yahballaha III (1281–1317), with the return of the Church of the East to China, even though its original territory in Persia had been considerably reduced in the preceding four centuries. We may note that an immense expansion of the Church of the East had already taken place in the time of the great patriarch Timothy I (780–823) at the beginning of the ninth century. Until the era of the patriarch Yahballaha III, during the ilkhanate (based principally in Persia), relations between the Mongol ilkhans and the patriarch of the Church of the East were positive. But the patriarch Yahballaha III himself witnessed the beginning of the decimation of his Church. Subsequently, the Christianity of the Church of the East, which had been able to maintain itself in Persia, Central Asia and China, scarcely survived the fall of the Mongol Empire. The destruction of the Church of the East in Asia towards the end of the fourteenth century is explained principally by the following facts: the fall of the Mongol Yuan dynasty in China, the aggressive Islamization pursued in Jagatai's khanate, and above all the persecutions inflicted by Tamerlane. Also the Black Death that swept through Central Asia played an important role. Together, these events marked the end of several glorious centuries in the history of the Church of the East.

For a time Syriac Christians believed that the Mongols would give them an enduring freedom. In the second half of the thirteenth century the important centres of the Church of the East were Mosul and Erbil together with the towns of Maragha, Urmia and Tabriz, which enjoyed a period of prosperity under the Mongol ilkhans. In the fourteenth century in Central Asia and Mongolia not much is known about the Christians of the Church of the East because the archives were destroyed when Christianity was eradicated in all these regions. We do have some indications, thanks notably to the written testimonies of several travellers and missionaries such as Giovanni dal Piano dei Carpini, William of Rubruck and Marco Polo, as well as the Armenian Smbat called Sempad the High Constable. The last great lists of the dioceses of the Church of the East that we have date from the end of the reign of Oljaytu and the beginning of that of Abu Said, in around 1316. Very few documents have likewise been preserved from the years 1316 to 1552. In the middle of the fifteenth century of the Church of the East had under its jurisdiction twenty-five ecclesiastical provinces. The province of Damascus consisted of six episcopal sees, two of which were the towns of Tarsus and Malatya (both today in south-east Turkey). By the sixteenth century in the exterior provinces close to the Mediterranean Sea, the Church of the East's communities had disappeared in Palestine and Syria, but the community in Cyprus continued to survive.

With regard to relations with Islam, it is true that for centuries Christians coexisted with Muslims often quite peacefully. We may recall that they had even collaborated on the work of translations that permitted the transmission of ancient Greek learning to the Arabs, participating together in the furthering of medicine, astronomy and other sciences (see Chapter 1). Nevertheless, this common feeling remained precarious, as has been explained.

In modern Iraq, after Tamerlane's capture of Baghdad in 1393, Christianity disappeared almost entirely from the city and the entire south of the country. In the north of modern Iraq the destruction of Erbil in 1397 by Tamerlane's forces delivered a fatal blow. The towns of Mosul and Gazarta, however, survived because they surrendered to Tamerlane. As Tamerlane's troops did not enter the area north of Mosul, this accounts for the Christians of the Church of the East seeking refuge in this region. It was there that a very small portion of the Christians of the Church of the East were able to survive. And it was to this area that the patriarchal seat was moved, to places that then seemed secure, to Kochanes in the high mountains of Hakkari (between the towns of Dohuk in northern Iraq and Van in eastern Turkey), for this region escaped the massacres and destruction at that time.

To conclude, the Church of the East was at that time almost completely wiped out. It disappeared from modern Iraq (except for the north of the country) and for the most part from Persia, as well as from Central Asia, Mongolia and China. It was reduced, as it were, to a territory situated to the north-east of the Euphrates, more or less in the north of its geographical region of origin in Mesopotamia. Consequently, the Church of the East survived principally only in two regions of the Middle East: in the mountains of the modern Turkish province of Hakkari and the surrounding area (a region situated partly in south-east Turkey and partly in northern Irak) and in neighbouring Persian Azerbaijan (in the north-west of modern Iran); and also in Kerala in India, but there too with great difficulty.

CHAPTER 7

The Nineteenth Century

Following successive persecutions, the most violent being that of the era of Timur Lang (Tamerlane) in the fourteenth century, the faithful of the Church of the East were obliged to seek refuge in north-eastern Mesopotamia, in the region lying between Mosul (in northern Iraq) and Lake Van (in south-east Turkey). In the nineteenth century they were therefore to be found established partly on Ottoman territory and partly on Persian. In the Ottoman Empire the Assyrian populations extended from the town of Amadiya (in the south) to the environs of the town of Van (in the north). They were mingled with a large Kurdish majority and some Armenians. In the Persian Safavid Empire, the Assyrians lived in Azerbaijan (today in north-west Iran), especially at Urmia, their principal centre, where they formed a good proportion of the population, and in the surrounding area. They also lived in the highlands of Tergavar, Mergavar and Baranduz (to the south of Urmia). There were still some Assyrians at Maragha, the city which the Mongol khan Hulagu (d. 1265) had made his capital on the eastern shore of Lake Urmia. We should not forget that there was also a small community of Christians originating from the Church of the East in Kerala (in south-west India).

In the mountainous region of Hakkari (in the south-east of modern Turkey) most of the villages where the faithful of the Church of the East lived were along the valleys of the Great Zab and its tributaries (the plain of Gawar, to the east and south of Kochanes) and in the region of Jilu (in the south-west of Gawar).

In this period the reason why the majority of the Christians of the Church of the East inhabited the Hakkari region and its environs was because their patriarch lived there and had his seat at Kochanes (20 km north of the modern town of Hakkari, the prefecture of the Turkish province of the same name, whose Kurdish name is Julamerk), itself situated 205 km from the town of Van and 173 km from Urmia. The seat of the patriarchs of the so-called Shimun line remained there from the seventeenth century until 1915, the year in which the Assyrian genocide forced the patriarch and his flock to seek refuge first in Persia (modern Iran) and then in modern Iraq.

It is probable that most of the 250 or so Assyrian villages of the Hakkari region that existed in the nineteenth century were founded in the last decade of the fourteenth century at the time of the flight into the mountains of the faithful coming from the south who settled there near communities of the Church of the East already established there much earlier. We also have archaeological evidence proving the presence of Christians of the Church of the East in this region from long before, notably churches that have been found dating from the pre-Islamic period.

As these communities of the Church of the East lived in remote places often difficult of access, it might be imagined that they were finally able to lead a peaceful existence, forgotten by the whole world. Even in these isolated retreats, however, the Assyrians were victims of local political disturbances, sometimes with international repercussions, as will be explained in this and the following chapter. In this period the region was the theatre of constant warfare between two neighbouring political great powers: the Ottomans and the Persians. To the east, the Russian Empire represented the greatest danger to both the Ottoman and the Persian governments since the time of its annexation of the Caucasus, which began with Georgia in 1801.

The Organization of the Assyrian Tribes

In 1850 in the region around Hakkari there were about 40,000 Assyrians grouped in villages organized in a tribal manner. Each tribe was ruled by a leader (called a *malek*), who governed locally. There were two groups of Assyrians: the larger was formed by the fairly autonomous tribes (*ashiret*) that lived in the central mountains of the districts of Tiyari, Tkhoma and the environs; the other was formed of those who lived in the lowlands and were subjects or vassals (*rayats*) more submissive to the Ottoman government and the local Kurdish chiefs (*agas*).

There were five main tribes: the Tiyari, the Tkhoma, the Jelo, the Baz and the Dez. In her book, Lady Surma, aunt of the patriarch Shimun XXI Eshai (1920–75), adds that of the Ishtazin. There are also other names to be added. The small tribe of the Dez (near Kochanes) was charged with the protection of the patriarch. The large tribe of the Jelo had its centre at Matha of Mar Zaya. To the west of the Jelo, the most numerous tribe was that of the Tiyari, which included about half of the Assyrians of Hakkari, and was centred on the village of Chamba of Malik. There were also other fairly large communities in the districts around Kochanes, at Albaq, Norduz and Van (to the north of Kochanes), and also in the Lewun valley, in Qaimar and Bohtan (to the west of Kochanes), and in the plain of Gawar and in Shemsdin (to the south-east of Kochanes). Around 30,000 Assyrians lived to the east of the Hakkari region, in Persia, in the plains of Urmia and Sulduz (on the western and southern shores of Lake Urmia). In the plain of Urmia there were about 150 villages inhabited by Assyrians, of which nearly fifty villages were in the Anzel district (to the north of Urmia) with the rest at Urmia, and about seventy in the district of Baranduz (near Urmia, to the south). About 3,000 Assyrians lived in the town of Eshnuq in the south of the plain of Urmia and in twenty-three villages in the neighbouring district of Sulduz. It is estimated that in 1877 approximately 80,000 to 100,000 Assyrians lived in these regions, distributed among 425 villages.

The Organization of the Church of the East

The patriarch resided at Kochanes, which lay in the district (*sanjak*) of Hakkari and was at that time part of the province (*vilayet*) of Van. It was a village perched at an altitude of 2,133 metres and surrounded by ravines and torrents flowing out to the Great Zab River, a tributary of the Tigris. The village today is in ruins. The only building left standing is the patriarchal church constructed in 1689 and dedicated to St Shallita. Seven patriarchs were buried to the west of the church. Today the town and the region of Hakkari are inhabited mainly by Kurds.

The patriarch was (and still is) regarded by all the faithful as the spiritual head of the Church. With regard to temporal power, the patriarch also represented the whole community in the Ottoman Empire, as a kind of civil magistrate and head of the nation. For example, it was he who had to settle conflicts, including those with neighbouring Kurds, in the latter case in the presence of the Kurdish emir, who exercised a higher authority. According to George Percy Badger (1815–88), this system permitted a kind of Assyrian state that was almost independent, while at the same time seeking a balance in its relations with the Ottoman authorities and the Kurdish emirs.

In the nineteenth century there were about fourteen dioceses in the patriarchate of Kochanes, in a zone lying between Bohtan (to the west of Kochanes, near the town of Siirt, today in Turkey), Berwari (near the town of Amadiya, today in northern Iraq) and the regions around Hakkari and further east (in Persia). All the dioceses at that time were situated either in the Ottoman Empire or in the Persian Empire. A little before 1890 there were twelve bishops of the Church of the East, nine in the Hakkari region and three in the plain of Urmia.

The Kurds

According to Lady Surma, relations between the Assyrian Christians and their Kurdish neighbours were often good on the whole, but there were also problems and conflicts. In the 1830s the leaders of the Kurdish tribes of this region (today called Kurdistan by some) managed to create a sort of Kurdish confederation that was almost independent of the Ottomans under the leadership of Badr, the chief (*emir*) of Bohtan (to the south of Lake Van). In 1841 Nurallah, the Kurdish emir of Hakkari, burned down the patriarch's residence in Kochanes. In 1843 Badr and Nurallah attacked the Christians of the Tiyari region. George Percy

Badger, an Anglican missionary of British nationality who was present in Mesopotamia and Kurdistan in 1842–4 and subsequently in 1850, recounts these massacres. A direct witness, he knew the situation of the Church of the East and its prelates very well and wrote an important book on the subject, *The Nestorians and their Rituals*. In a letter addressed to him on 4 June 1843 by the patriarch Shimun, the patriarch gives an account of the events. Many women and children were taken and led into captivity, and some of them were sold as slaves. In the Dez region the Kurds cut in two the body of the patriarch's mother, an elderly woman, and threw it into the River Zab. Numerous people were killed (more than 3,000), among them the leaders of the Tiyari, and some ecclesiastics. The Kurds burned and destroyed entire villages. No churches survived and only a few houses remained standing. These events were confirmed in other letters of the patriarch. For example, in a letter which he sent to the Anglican bishop of London on 12 August 1843 the patriarch writes that the Muslims of Kurdistan had eliminated the Christians almost entirely in the provinces of Tiyari and Dez. This explains why, in 1843, the patriarch fled to Mosul and sought asylum at the British consulate. A few hundred of the faithful followed him. According to Badger, they could not return to their homes for fear of the Kurds. Then the patriarch fled to Urmia in Persia, where his faithful were fairly numerous. Sir Stratford Canning, who for a long time was the British ambassador to the Ottoman Empire, intervened to plead the case of the unfortunate Assyrians at the Porte, that is to say, with the Ottoman government at Constantinople (called Istanbul after 1930), and the British vice-consul at Mosul, Christian Rassan, did the same. There were also diplomatic representations on the part of Russia.

In September 1846 a new massacre of the Assyrians took place when the Kurds, with the emir Badr at their head, sacked the villages in the district of Lower Tiyari (around the village of Ashita) and Tkhoma (to the east) and massacred the villagers. The survivors fled to Persia. According to Badger, all the villages and churches were destroyed. Following these dramatic events the Ottoman government was called on by the foreign diplomatic representations to put an end to the power of the Kurdish emirs. The emir Badr and his ally the emir Nurallah were then sent into exile. In 1847 the Ottoman government sent an army to central Kurdistan to suppress the Kurdish emirates and put in place a direct Ottoman administration. This was fully established in 1850. When calm had been restored, the patriarch Shimun was finally able to return to Kochanes after an absence of several years. No compensation was offered to the Church of the East and its people.

With all the difficulties that they had to face, the Assyrians understood that they needed the protection of the great powers. This explains why they came into contact first with the Russian Empire, which was the Christian power nearest geographically to them and which was present in Transcaucasia (Georgia, Armenia) and the whole of the north of Azerbaijan in Persia (including the shores of Lake Urmia) following the treaty of Turkmenchay signed in 1828. In 1868 the Assyrian patriarch contacted the Russian tsar's brother to ask him for assistance. In 1877 the Russians invaded a part of eastern Anatolia (modern eastern Turkey). At the end of the nineteenth century the Russians pursued their plan of conquest in the direction of Persia. The Assyrian leaders also placed great hope on the British. Consequently, towards the end of the nineteenth century the Church of the East found itself caught between the rivalries of the great powers of the period, the Russian Empire, Great Britain and even France, which all sought to impose their influence on the Ottoman Empire. By the beginning of the twentieth century the Assyrian Christians had become a kind of plaything in the political strategy of the great powers just mentioned and their respective Churches (Orthodox, Anglican and Catholic).

To all these political and religious intrigues may be added other difficulties, such changes to the frontiers. Following its policy of centralization, in the 1840s the Ottoman government incorporated the regions of Bohtan (around the modern town of Cizre in Turkey) and Bahdinah (around the modern town of Amadiya, today in the province of Dohuk in Iraqi Kurdistan) in the district (*sanjak*) of Mosul. Ottoman authority also extended as far as Jezireh (today in north-east Syria) and to the frontiers of the Tiyari country, which allowed the Ottoman government better control over the region called Kurdistan.

The Missionaries

In the nineteenth century the missionaries of various Churches who came from abroad also played a role in the history of the Church of the East. From the 1830s Protestant missionaries, mostly Americans and British, began to work among the faithful of the Assyrian patriarchate of Kochanes, first in the regions of Urmia (Persia) and also Hakkari. In the 1830s the success of Catholicism at Mosul was due to the work of missionaries. The Catholic mission also developed in Persia.

The earliest mission was that of the American Presbyterian Church. Its first missionary was Justin Perkins (d. 1869), who established a missionary centre at Urmia in 1835 with Asahel Grant of the American Board of Commissioners for Foreign Missions, a mission transferred in 1870 to the Presbyterian Church, which in 1874 established a Reformed Church at Urmia (numbering about 2,700 members in 1914).

The Anglicans began their missionary activities in 1843 when the then archbishop of Canterbury sent a mission consisting of one priest, Fr George Percy Badger, who has already been mentioned. In 1876 Fr Cutts was sent in turn to study the situation on the spot. In 1881 the Anglican mission of the archbishop of Canterbury was established, which was active until 1904. In 1886 the Anglican mission was reorganized by three priests, Frs Riley, Maclean and Browne, the last of whom lived for several years at Kochanes where he died in 1904. It seems that the Anglicans were keen that the Assyrians should maintain their beliefs and ancestral customs. They wanted above all to aid them, for example, by establishing schools and translating and publishing the books of the Church of the East.

As for the Catholic missions, their activities began in the sixteenth century, but their work was only really organized towards the middle of the eighteenth century, when the Dominicans arrived at Mosul. By the end of the nineteenth century very few members of the Church of the East remained at Mosul and in the surrounding plain, an area swamped by a majority of Chaldean Catholics. Alqosh, near the monastery of Rabban Hormizd, which had become Catholic in 1808, became a Catholic bastion. In the region of Erbil the last Assyrian villages (Ainqawa, Armuta and

Shaqlawa) were converted to Catholicism towards the end of the eighteenth century. At Baghdad the Chaldean Catholic community grew during the nineteenth century. In Persia, Assyrians went over to Catholicism especially at Khosrowa since 1863 and at other villages in the plain of Salmas since the end of the eighteenth century. The same took place in the large villages of the district of Tergawar (to the south of Urmia), in the second half of the nineteenth century. In 1844 the Ottoman authorities issued a decree (*firman*) officially recognizing the Chaldean Catholic patriarch Nicholas I Zaya (1840–7) and his Church. This was made possible by the intervention of the French ambassador at Constantinople. We may note here that the Church of the East was never officially recognized by the Ottoman government as constituting a community (*millet*). By 1830 about one third of the total population of the Church of the East had become Catholic. The Catholics hoped that all the faithful of the Church of the East would join the Catholic Church. The Chaldean Catholic patriarch Elias XII Abulyonan (1879–94) sent missionaries to the distant Assyrian villages of Hakkari, who penetrated the region by going up the Great Zab River. Moreover, in the last part of the nineteenth-century Latin Catholic missionaries came and established themselves in nearly all the regions where there were Chaldean Catholics: the Dominicans at Mosul, the Carmelites at Baghdad and the Lazarists in Persia. All this explains why the Chaldean Catholic community grew so rapidly and attained the size it did. In 1961 John Joseph wrote: 'Most probably the Catholic Church will entirely absorb the remnants of the Nestorian Church which to its faithful few is known by its proud but once-fitting name "The Old Church of the East".'

The Church of the East also had contacts with the Russian Orthodox Church. Since 1851 the Assyrian bishop Yosip of Ada (to the north of Urmia) had sent a representative to Erevan (Armenia) and Tiflis (Georgia), under Russian occupation. In 1861 an Assyrian priest from Urmia travelled to St Petersburg. The Holy Synod of the Russian Orthodox Church then sent a bishop who came to study the situation of the Church of the East on the spot. In April 1868 the patriarch of the Church of the East sent a letter to the Russian governor-general of the Caucasus to inform him of the very difficult situation of his faithful, about 16,000 families, who lived in the mountains of Kurdistan. In 1884 Bishop Gabriel of the Church of the East went to Tiflis and negotiated a project of union with the Russian Orthodox Church. The Assyrian bishop Yonan of Sopurghan (to the north of Urmia) was the last bishop of his Church in the plain of Urmia. In 1898 he went with several members of his clergy to St Petersburg and they became Orthodox. In the same year the Russian Church sent a delegation of several priests and monks. The Russian mission was installed in Urmia in 1898. In 1900 a church was erected in Urmia and some parishes and schools were also established. There were two bishops of Assyrian origin who were appointed by the Russian Orthodox Church, Yonan in 1898 (d. 1911) for Sopurghan and in 1904 Elias Abraham (d. 1928), Yonan's vicar, for Tergawar. By 1914 the Russian mission was well established. So in 1914 a Russian bishop, Sergius Lavrov, was sent to Urmia. He was joined by another Russian bishop, Pimen Belolikov, who was ordained for Salmas in 1916. Some 10,000 to 20,000 Assyrians were converted to Russian Orthodoxy in this period. The Russian Orthodox mission founded a press and a journal: *Urmi Artuduksita* (Orthodox Urmia) printed in Russian and Assyrian. In 1906 a liturgical book was published in the classical Syriac language. In addition, emissaries of the patriarch of the Church of the East went to Tiflis to administer the affairs of the Assyrian community that resided there.

Moreover, in the nineteenth-century context that we are discussing, the competition between the missionaries, especially Protestants and Catholics, for Assyrian converts must be emphasized. All these missionary activities provoked cultural and other changes as well as additional divisions among Assyian people and within the Church of the East.

In the nineteenth century, as in the preceding centuries, we have direct testimonies relevant to the history of the Church of the East in the writings of a number of missionaries, travellers and archaeologists, among whom are Ainsworth, Cutts, Badger, Layard, Isabella Bird Bishop, Maclean and Browne, Perkin, Fletcher, Wigram and also Shedd.

CHAPTER 8

The Twentieth Century

The Beginning of the Twentieth Century

At the beginning of the twentieth century the geographical area where the Christians of the Church of the East lived was limited principally to the mountain massif of Hakkari, situated today on the frontiers of south-east Turkey and northern Iraq. Apart from this area, leaving aside India, only the regions of Urmia (in the Persian Empire) and of Amadiya and Van and their environs (in the Ottoman Empire) held a substantial number of Assyrian Christians. In the Middle East in 1913, that is to say, on the eve of the First World War, their numbers are reckoned as follows: a little over 60,000 in the Ottoman Empire and 30,000 in the Persian Empire, a total of between 100,000 and 120,000 people, at that time still a little higher than the number of Christians of the Chaldean Catholic Church, who are estimated to have been about 100,000 in 1913. Just before the First World War the ecclesiastical centre of the Church of the East was still situated in the village of Kochanes (Hakkari), where the residence of the patriarch lay, on the south-east frontier of the Ottoman Empire (modern Turkey). On the eve of the First World War there were at most eight bishops of the Church of the East.

With regard to the internal politics of the Ottoman Empire, the beginning of the twentieth century was a disastrous period which saw the greatest catastrophe experienced by the Church of the East since the ravages of Tamerlane in the fourteenth century, as we shall now see. Under the Ottoman sultan Abdul Hamid II (1876–1909) many Christians were massacred, including Assyrians. The sultan was deposed as a result of the Young Turk revolution in 1908. The Young Turks very quickly disappointed those who at first had trusted in their promises of political and democratic reforms. In fact their ultra-nationalist policy degenerated into a pogrom against non-Turkish minorities.

In 1915, on the 24 April (the commemoration day of the Armenian genocide), the minister of the interior Talaat Pasha of the Young Turks' government gave the order for the arrest of Armenian intellectuals, with deportations and massacres to follow. In September 1915 the minister ordered the extermination of all the Armenians. It is known that the Armenians were victims of massacres in 1894 and 1909, of deportations in 1915 and of genocide in 1915–16. The Christians of the West-Syriac and East Syriac traditions, to which the Christians of the Church of the East belonged, suffered the same fate. In this period the total number of Armenian and Assyro-Chaldean victims numbered more than two million.

With regard to international politics during the First World War, the Caucasus and eastern Anatolia was one of the regions that experienced the full brunt of the war because it lay on the very strategic Russian-Ottoman front. This front passed through north-west Persia (modern Iran) and its adjoining areas, a zone that had been the object of intense political competition between the Ottoman and Russian Empires, with alternating occupation by them. In fact from 1914, with the outbreak of the First World War, the Ottomans and the Russians were engaged in constant fighting with each other. For the Ottomans the principal threat was a Russian invasion of their eastern provinces, regions also inhabited by Christians. During the First World War, the Assyrians who lived in this war-torn zone again suffered greatly from the disasters that overtook them as a result of political circumstances completely beyond their control. Here now is a summary of what they experienced and the exiles that followed.

History and Exiles (1914–18)

The Year 1914

It was in November 1914 that the Ottoman Empire formally entered the war, provoking anxiety among the Assyro-Chaldean Christians for their future. Christian minorities in different parts of the Ottoman Empire had had experience at that time of massacres in their communities. During the second half of 1914 the patriarch Shimun XIX Benyamin (1903–18) received the guarantee of the Ottoman government and of the governor of the town of Van that his people had nothing to fear. According to Lady Surma (1883–1975), of whom more will be said later, on the 3 August 1914 the patriarch was summoned to Van to meet the Ottoman *vali*, Tahsi Pasha. They had a long conversation about the assistance which would be offered to the Assyrians provided that they remained neutral and did not join the Russians. The patriarch's response was clear. He said that the attitude of the Assyrians would depend on that of the Ottomans towards the Christians in general, and that with regard to the matter in hand the patriarch had necessarily to consult the leaders (*maleks*) and notables of his nation because he could not speak simply on his own. The patriarch returned from Van and sent letters to all the districts where he had members of his community, urging them to remain calm and asking them to fulfil all their obligations towards the Ottomans. But during the winter of 1914 Turkish offensives were launched against the Assyrian villages in the regions of Albaq and Gawar, which were attacked and pillaged. The promises of the Ottomans proved to be empty.

The Years 1915–17

The year 1915 was the worst of all. The Assyro-Chaldeans remember it as 'the Year of the Sword' (*shanta d'Sayfa*) because many members of the Christian minorities were massacred in that year in several eastern provinces of the empire by the Ottomans and their Kurdish auxiliaries. In Hakkari, which was a more isolated region, the Assyrians were at first not affected by these events.

In Persia in January 1915 the Russians withdrew their forces, and the Ottomans occupied Tabriz and a large part of Persian Azerbaijan, taking the town of Urmia on 2 January 1915. The Assyrians thus found themselves at the mercy of the Ottoman forces. During the Ottoman occupation of Persian Azerbaijan, Christians were massacred and very many villages (about three quarters of them, it is said) were burned and sacked. When the Assyro-Chaldeans learned of the Russian retreat, thousands of them followed the Russian troops northwards to seek refuge in Transcaucasia. The same happened each time the Russians were forced to retreat, which explains the Assyrian migrations to this zone, at that time in Russian hands.

Further west in the Ottoman Empire, at the beginning of 1915 the Russians mounted a military offensive in the direction of the towns of Erzerum and Van, and the Ottomans retreated to Siirt. In May 1915 the town of Van, near Hakkari, was taken by the Russians. At the end of April 1915 Russian troops penetrated the high part of the Great Zab valley, in the district of Albaq. The Assyrians of Hakkari then had to make a choice: to maintain their neutrality towards the Ottomans, or to join the Russians, who at that moment represented a hope of liberation from the Ottoman yoke. At the beginning of 1915 contacts between Assyrians and Russians were intensified.

As David Gaunt explains, the Assyrians were exasperated by repeated assaults and massacres, perpetrated against their community by the Ottomans and the Kurds. Moreover, the first reports of massacres of Armenians were terrifying. A council of leaders (*maleks*) of the Assyrian tribes took the decision to declare war on the Ottomans on the 10 May 1915. This decision was of course disastrous, given that the Assyrians were vastly inferior in numbers. It triggered the fury of the Ottomans and in June 1915 Ottoman troops attacked the Assyrian villages of Hakkari. As a reprisal, the patriarch brother, who was held in Mosul, was executed by the Ottoman authorities in July 1915. In the summer of 1915 regular Ottoman troops accompanied by Kurdish forces advanced against the Assyrian settlements of Hakkari, including Kochanes, the patriarchal seat. The Assyrians found themselves encircled and forced to abandon their villages and retire higher up the mountain to their summer pastures. The villages and churches of the Church of the East were pillaged and destroyed. The situation had become desperate. In October 1915 there was no other course of action but to flee to Persia, where there was quite a large Assyrian community. There the Russian authorities organized assistance for the refuges in the districts of Khoy, Salmas and Urmia. The missionaries also offered help. It was thus that the entire Assyrian population of

Hakkari, men, women and children with their patriarch left the land of their birth. This exodus led to the disappearance of the Assyrian people from their mountains of Hakkari. By the time they reached the zones occupied by the Russians in Persia it appears that almost half of the Assyrians had perished from massacre or disease.

In December 1915 the patriarch Shimun XIX went to Tiflis (the capital of Georgia, today called Tbilissi) in order to inform the Russian military authorities of the desperate situation of his people and ask them as a matter of urgency for aid and protection. About 15,000 Assyrians were thus authorized to come and settle in Armenia and Georgia.

In 1916 the Russians took the towns of Erzerum and Erzincan (both in the north-east of the Ottoman Empire) and the Ottomans stationed in Persia were obliged to withdraw. But in March 1917 Tsar Nicholas II abdicated. The revolution of October 1917 a little later marked the beginning of the communist period in Russia. Russian troops at that time had to leave the Urmia region and the Ottoman Empire. This resulted in the Assyrians finding themselves isolated and without support. Other terrible trials now awaited them.

In Persia in 1918

In Persia, in February 1918, the Muslims of Urmia rose up against the Christians. In the spring of 1918 the Ottomans penetrated Persian Azerbaijan. On the 18 July 1918 they took the town of Urmia and massacred thousands of Christians. It was then that the Assyrians, together with other local Christians, decided to leave the town of Urmia and its environs. Numerous Christians, among them the Assyrians of Hakkari and Urmia, made their way on foot southwards as far as Hamadan (536 km south of Urmia) in the hope of finding help from the British. The town of Hamadan had been taken very recently by the British army, who had already conquered the Iraqi capital, Baghdad, in March 1917. According to Lady Surma and David Wilmshurst, about 50,000 Assyrians of the 70,000 who left Urmia survived the march.

The Assassination of the Patriarch Shimun XIX Benyamin. The New Patriarch. Lady Surma

Before this drama, the Assyrian Church and the Assyrian community were victims of another catastrophe in March 1918, when the patriarch Shimun XIX Benyamin (1903–18) was treacherously assassinated. On the 3 March the head (*aga*) of the Kurdish tribe of the Shikak, Ismail Simko, organized a meeting with the patriarch Shimun XIX Benyamin at his residence in the village of Kohnashar (near Salmas). The patriarch announced that he would go there in order to discuss peace. But at the rendezvous an ambush had been prepared. The patriarch and some fifty-five men of his escort were killed by a fusillade as they left the meeting. The body of the patriarch, who is considered a martyr by the Assyrian faithful, was taken to Khosrowa, where it was buried in an Armenian Church. The Assyrians had lost their spiritual and temporal leader and were now more isolated than ever in the midst of a hostile environment.

On the 29 March 1918 a new patriarch was enthroned, the youngest brother of Shimum XIX, under the name 'Shimun XX Paul' (1918–20), in the ancient Church of Mart Mariam in Urmia. But Shimun XX, exhausted by the march into Iraq and all kinds of privations, and suffering from tuberculosis, died very soon after. He was thirty years old. It is said that he was buried in the courtyard of the Apostolic Armenian Church in Baghdad. The only member of the patriarchal family who was fit to succeed him was his nephew Eshai, the future patriarch Shimun XXI Eshai (1920–75). Born on the 26 February 1908 at Kochanes, he was sent to England in 1924, where he studied at Canterbury and Cambridge for three years.

Because of the new patriarch's extreme youth, his aunt, Lady Surma (1883–1975), the niece of the patriarch Shimun XVIII Rubil (1861–1903), acted as a kind of regent for the Church of the East. She was also the sister of two patriarchs: Mar Benyamin (1903–18) and Mar Paulos (1918–20). Having received a good education through the intermediary of Anglican missionaries at Kochanes, in particular the Rev. W. H. Browne, she was very active in helping her Church in different ways and in various circumstances.

She notably maintained numerous contacts among highly placed ecclesiastical and political figures, particularly in England. She was invited by the British authorities to present the 'Assyrian question', of which we shall speak below. It was thus that in October 1919 Lady Surma arrived in London to plead the cause of the Assyrian people as a kind of ambassador for her community. The matter was debated in the House of Lords on 17 December 1919. Lord Curzon put forward the project of an Assyrian enclave in the north of the *vilayet* of Mosul, under British control. In February 1920 Lady Surma drew attention to the rights of her people and the fact that Great Britain had recognized the Assyrians as allies. After the conference at San Remo (April 1920), Lady Surma wrote that she judged a mandate under British protection in the Hakkari district preferable for her people, while realizing that this was impossible. The alternative therefore would be either to live in Hakkari if an independent Kurdistan was established there, or to remain in Iraq under British mandate in a region offered to the Assyrian nation near Altun Kupri (north of Kirkuk in northern Iraq). In November 1925 Lady Surma went to Geneva to testify to the situation of her people. On the 8 December 1925, during the thirty-seventh session of the League of Nations, she went to the rostrum to speak of the pitiable state of the Assyrian refugees. Lady Surma was thus able to make the problems of her community known at the highest level. She did not want her people to be dispersed because that would prove fatal for them. But this people only represented a small community without interest to the great powers, which did not give it all the attention it needed. Lady Surma was a direct witness of the exodus of her people, which she describes in her book *The Assyrian Church Customs and the Murder of Mar Shimun*. According to what she says, many more books would be needed to recount all the acts of oppression suffered by her people.

Among the other direct witnesses to the atrocities committed which we are recalling, we should mention missionaries and diplomats who have left us their written testimonies on this subject, for example, those of two consuls at Urmia, Basil Nikitin (of Russia) and William Ambrose Shedd (of the United States), who was a former missionary. For a detailed history, the British, French and American state archives should be consulted as well as the international press of the time.

In Iraq from 1918

As has already been explained, it is because the Christians understood that there was no longer any security for them if they remained in Persia that they undertook their desperate walk to reach the British forces at Hamadan. There, they were given some emergency help. From Hamadan they were moved to Iraq, to a refugee camp at Bakuba (68 km north-east of Baghdad). The route from Hamadan to Bakuba was itself also long (508 km). Later, other camps were established for them at Habbaniya (94 km to the west of Baghdad) and elsewhere in the north of the country.

In 1921 there were no more than four bishops in the Church of the East: Joseph (Yosip) Khnanisho, Zaya Sargis, bishop of Jelo; Baz and Raikan, Yalda Yahballaha, bishop of Berwari; and Abimalek Timothy, metropolitan of Malabar in India.

In 1922, when the Assyrians who had come from the Urmia region returned to their villages, they were only a fifth of the number they had been in 1914. In 1966 it is estimated that there were about 1,400 families living in the larger villages of the Plain of Urmia. Some Assyrians succeeded in returning to their villages in Hakkari, but in September 1924 the Turkish army drove them out and burned the villages that they had reoccupied. Some hundreds of them were killed and the survivors returned to Iraq. For them there was no other choice but to establish themselves in Iraq. After 1925 the majority of Assyrians originally from the Hakkari region were resettled in villages in the north of Iraq.

The Simele Massacre (1933). The Patriarch Driven Out of Iraq

In Iraq the kingdom (1921–58) that had been placed under British mandate obtained independence in 1930, although British tutelage remained strong and the British retained some military bases. The question of the status of the Assyrians was not resolved. In 1925 the League of Nations had reattached Hakkari to the new Turkish Republic and had forbidden the Assyrians to return to their ancestral lands. The great majority therefore remained in Iraq.

In 1933 the patriarch Shimun XXI Eshai, who represented his people, asked for a certain autonomy to be granted to the Assyrians in Iraq (see 'The Assyrian Question', below). On the 28 May 1933, the minister of the interior at Baghdad, while recognizing the patriarch's spiritual authority, asked him to abandon his temporal authority. The patriarch refused and was placed under house arrest by the Iraqi government in June 1933.

A number of Assyrians decided to leave Iraq and settle in Syria, at that time under French mandate (1923–46). On the 21 July 1933 about six hundred Assyrians tried to cross the frontier to enter Syria, hoping to receive political asylum. But they were arrested by the French authorities and on the 4 August sent back across the northern Iraqi frontier, near Dohuk. On the road, as they returned, confrontations arose with Iraqi soldiers at the frontier that rapidly degenerated. This event was immediately followed by a veritable drama when new massacres were carried out from the 7 to the 15 August 1933 at Simele (16 km from Dohuk) and the surrounding area. On the 7 August 1933 there was an outbreak of gunfire at Dohuk and Zakho. From the 11 to the 15 August at Simele troops belonging to the Iraki army massacred Assyrian men, women and children. Massacres also took place at Dohuk, Zakho and elsewhere. About sixty Assyrian villages were pillaged and more or less destroyed in the mountains to the north of Simele. According to official British sources, the number of Assyrians killed is estimated to have been about 500, but other sources speak of 2,000 to 3,000 dead. Lieutenant-Colonel R. S. Stafford, who was at that time British administrative inspector at Mosul, described this event as 'a systematic massacre carried out in cold blood'. He adds that the spirit of the Assyrians was totally crushed. Indeed, these exactions and these killings inflicted a terrible toll, further weakening the Assyrian community, which had already been deeply ravaged by earlier events. In consequence, the 7 August is commemorated by the Assyrian community as a national day of mourning, 'the Day of the Martyrs', in memory of the Simele massacres. This was declared by the Assyrian Universal Alliance in 1970.

In Iraq the aggressors were considered heroes and were not punished or prosecuted. We should recall that in the 1920s thousands of Assyrian men were enrolled in the local British military forces called the Levies. It was consequently this that led to their becoming unpopular and held in contempt, for they were considered allies of a foreign colonial power, Great Britain, which had recently occupied the country. In 1955 the corps of Levies in Iraq was disbanded and most of the Assyrians Levies joined the Iraqi army.

As a result of this new massacre, a part of Iraq's Assyrian population decided to flee to the north-west of Syria, to the Khabur valley (which lay to the west of the town of Hassake, 250 km to the south-west of Dohuk), where they built about thirty villages. By the end of 1933 about 8,500 Assyrians had already settled there. Other Assyrians subsequently joined them. After the end of the French mandate in Syria (1923–46), they were authorized to stay.

In Iraq on 18 August 1933 the patriarch and his family were expelled by the government and exiled to Cyprus, with the loss of their Iraqi nationality. The members of the Church of the East now found themselves without a patriarch. The affairs of the Assyrian Church in Iraq were entrusted to Mar Joseph (Yosip) Khnanisho (1893–1977), metropolitan since 1918 of Shamsdin (in the ancient province of Adiabene), who lived at Harir (to the north of Erbil). In 1963 he transferred his episcopal seat to Baghdad. In his role as vicar-general he was also responsible for all the other dioceses of his Church in the Middle East. In the 1960s other bishops were ordained for the towns of Erbil, Kirkuk, Basra and Baghdad. In May 2019 Mar Yosip Khnanisho was canonized by the Holy Synod at Erbil; he is to be commemorated each year on the second Sunday of July.

With regard to the Assyrians who wished to leave Iraq, the League of Nations decided in September 1937 that they had to remain there as simple citizens. The Iraqi Ministry of Foreign Affairs declared that 'the Assyrian community could resume its place as an ordinary minority profiting from the benefits accorded by the declaration on the protection of minorities [...] while fully respecting their legal obligations vis-à-vis the Iraqi state'. In 1938 an Iraqi government report sent to the League of Nations speaks of seventy-three villages settled by Assyrians originally from Hakkari in the areas of Amadiya, Dohuk and Rawanduz (123 km from Erbil). Other Assyrians settled in large Iraqi towns, such as Baghdad, Kirkuk and Basra;

some still lived in camps. It is estimated that in the 1930s there were about 20,000 Assyrians in Iraq.

The first constitution of the Republic of Iraq dates from 1925. Although the Chaldean Catholics were allowed to be represented in the Senate by their patriarch, all national representation was forbidden to the Assyrians under the pretext that they were not nationals but refugees who had come from Turkey after the 1914–18 war. Latterly, and even until today, there have been Assyro-Chaldean representatives in the parliament at Baghdad and at Erbil in the regional government of Kurdistan in northern Iraq. Until 2008 the following quotas were assigned to the Assyro-Chaldeans in six governorates: three seats at Baghdad, two at Erbil, two at Kirkuk, two at Dohuk and one at Basra. Since 1992 in the regional governorate of Iraqi Kurdistan (RGK) five seats have been reserved for Christians (at that time 5% of the population) out of the one hundred and five seats of the Kurdish parliament.

Reconstruction of the Patriarchate in Exile

In consequence, a large number of Assyrians emigrated, mostly to the United States, where the patriarch Shimun XXI Eshai joined them in 1940 and where the Church was reorganized. The patriarchal seat was established in Chicago. In 1949 the patriarch received American nationality. In the United States he ordained a bishop for a newly created diocese of America and Canada. He also maintained contacts with his communities and bishops in the diaspora, including the Middle East, to whom he made visits. In 1968 the patriarch had his Iraqi nationality restored. By a decree dated 21 May 1970, the Iraqi president, Al-Bakr, recognized Mar Shimun XXI Eshai as patriarch of the Church of the East and supreme head of the Assyrian people in the republic of Iraq. In 1973 the patriarch Eshai resigned for personal reasons. With him the centuries-old organization of the hereditary patriarchate (that is to say, the transmission of the patriarchal title within the same family) came to an end. This kind of system may be explained by the fact that it allowed the patriarchal succession to be secured more directly and therefore in principle without conflict, thus ensuring the survival of the patriarchate. Even if some people criticize this system and consider it to have been a weakness, it may be admitted that there were advantages when all the difficulties endured by the Church of the East in the course of centuries are taken into consideration.

Schism within the Church of the East (1968)

In 1964 another very serious problem arose within the Church of the East when, following a synod, Patriarch Shimun in a synodical letter dated 28 March announced certain changes in the life of the Church. Among these was the use of the new calendar (the Gregorian calendar current in the West), the limitation of days of fasting, and the shortening of the text of the Liturgy. These novelties were considered too Western and modern by traditionalist believers. Although the majority of the community accepted these changes, some others were unwilling to do so, particularly in Iraq. The ecclesiastical figure at the head of the group of traditionalists who refused to accept the reforms introduced by Patriarch Shimun was Thoma Darmo, from 1952 metropolitan of Thrissur (in India). He opposed both the hereditary succession of patriarchs and the adoption of the Gregorian calendar. This small group declared its opposition through organizing the election of another patriarch.

In 1964 Mar Darmo was canonically suspended by Patriarch Shimun. In September 1968, however, Mar Darmo flew to Baghdad, where he ordained some priests together with three bishops: Mar Paulos (d. 1998) and Mar Aprem for India (with their seat at Trichur) and then Mar Addai for Iraq (with his seat at Baghdad). In October 1968 these three bishops enthroned Mar Thoma Darmo at Baghdad as patriarch of the so-called 'Ancient Church of the East' and deposed Patriarch Shimun, who in turn deposed all of them. This led to a patriarchal schism within the Church of the East and to the foundation of the Ancient Church of the East, a schism that has lasted to the present time despite repeated efforts to restore unity. Such was the beginning of the history of this Church, whose seat is at Baghdad. It created in consequence a rival double hierarchy within the mother-Church. After the death of Thoma Darmo in 1969, his successor, Addai, was enthroned in 1972. At the time of writing (2019) he was still in office.

It should be mentioned that behind this conflict lay the influence exerted by certain tribes – for example, the Tiyari tribe supported Mar Darmo, an Ashita. The present patriarchs, Mar Giwargis II (of the Church of the East) and Mar Addai II (of the Ancient Church of the East) belong to the same tribal group, the Ashita, of the Tiyari clan. The Iraqi authorities were also involved in this affair. In fact, the government at the time (the socialist Ba'ath party of the Arab renewal), made it known that it preferred a patriarch who resided in Iraq. In September 1968 a government decree assigned all the churches under the authority of Patriarch Shimun to the dissident group. But in December 1969 the Iraqi government invited the canonical patriarch to visit Iraq, which he did in 1970. This encouraged many of the faithful who were following the new calendar in Iraq. At that time the confiscated churches were restored.

Patriarch Dinkha IV (1976–2015)

In October 1976 the new patriarch, Dinkha IV Khanaya (1976–2015) was enthroned in Ealing (London). He was born in the village of Drerbenkul (near Erbil) and had been metropolitan of Teheran and Iran since 1962. Under Patriarch Dinkha the patriarchal seat remained at Chicago. In 1976, following the election of the new patriarch, Dinkha IV, the fact that the Holy Synod of the Church of the East decided to abolish the custom of the hereditary succession was a very positive step, for this was a critical matter for many Assyrians. In October 2005 exchanges took place between the patriarch and Masoud Barzani, president of the regional government of Kurdistan (in northern Iraq) to construct a patriarchal residence at Erbil, with the hope of establishing the seat of the patriarchate there. In September 2006 Patriarch Dinkha visited Iraq. In addition, the patriarch was able to heal the schism that had taken place in 1968 with the important metropolitanate in India.

The Patriarch Giwargis III (2015-)

In 2015 a new patriarch was elected, Giwargis III Sliwa. Born in 1941 in Habbaniya (Iraq), he had been enthronized in Chicago in 1981 as metropolitan of Baghdad and all Iraq. In 2014 he had made a visit to Turkey (to Tur Abdin and Hakkari), a visit all the more striking because it was the first visit by an Assyrian bishop since the genocide of 1915. His patriarchal enthronization took place at Erbil on 27 September 2015. Since that date the patriarchal seat has been moved from Chicago to Erbil. This very courageous decision of the Church of the East to be close once again to the faithful in Iraq and the Middle East, to sustain them in difficult political circumstances, often dangerous, and above all to encourage them to continue to live in their ancestral lands in the Middle East and not emigrate abroad, deserves praise.

The Assyrian Question

Among the uprooted Assyrian people there emerged a feeling of national identity, that is to say, the awareness of constituting an Assyrian nation. The Assyrians sought to be treated as a homogeneous group with administrative autonomy and an autonomous enclave, as had been promised to them by the British in 1917. In 1915 some Assyrians founded in the United States an early national movement, the Assyrian National Association of America, an association directed by Joel E. Warda (who was close to the Assyrian patriarch). At the Peace Conference in Paris (1919) they deposited a memorandum asking for the creation of a separate Assyrian state, supervised by a mandatory state. The Assyrians of the Caucasus were active too. In 1919 various Assyro-Chaldean delegations also came to the Peace Conference in Paris. These communities made their appearance at that time on the international scene. Some prelates also went to the conference, but the Assyrian patriarch Shimun XX was unable to do so. Before his death in 1920, however, he was of the opinion that the Assyrian refugees should be permitted to return to Hakkari and the Urmia region under a British protectorate.

In 1920 the new patriarch, Shimun XXI Eshai (1920–75), made representations that his people in the regions of Hakkari and Urmia should be enabled to live together in the region of Mosul and Amadiya under a British protectorate. In 1920, at the Treaty of Sèvres, the autonomy of some minorities was discussed and the possibility was envisaged for establishing independent states for the Armenians, the Kurds and the religious minorities, including the Assyro-Chaldeans. The term 'Assyro-Chaldean' appears in the treaty. In 1923, however, at the Treaty of Lausanne, there was no longer any question of these minorities.

In 1924 the Assyrians and Chaldeans made an appeal to the League of Nations. In 1932, at the end of the British mandate in Iraq, the problem of the resettlement and repatriation of the Assyrians was still unresolved. In the same year the patriarch Shimun went to Geneva to meet officials of the League of Nations at a conference held in December and asked for autonomy for the Assyrians, but without success. In 1933, after the Semele massacre of August, the League of Nations commissioned a study to find out whether the Assyrians could be resettled in Syria, but in 1936 the project was abandoned. On 16 August 1936 Patriarch Shimun sent a petition to the League. He then went to Geneva in person on 4 October 1933 and the Assyrian Question was examined on the 13 October. On the 24 October the patriarch submitted some suggestions to the League about the plight of his people and the possibility of moving them out of Iraq. In 1933 the League of Nations rejected the request of the Assyro-Chaldeans to be recognized as a homogeneous group who should be accorded an autonomous region. In September 1937 the League of Nations considered the Assyrian Question closed. Mar Shimun immediately went to Geneva but was unable to achieve anything. The Assyrian Question was at that time closed definitively so far as the League of Nations was concerned.

In a letter of the patriarch Shimun XXI Eshai addressed to the archbishop of Canterbury, dated 19 June 1943, the patriarch wrote that in 1937 the League of Nations, in full accord with the British government, had turned its back on the Assyrian Question. According to Claire Weibel, the Assyrians had been made scapegoats of Iraqi politics and the promises made by the British for an autonomous territory were not honoured: the hopes of the Assyro-Chaldeans had been disappointed. Once again promises were broken and the world forgot the Assyrians. It was in vain that in 1945 and 1947 the patriarch (Mar Shimun XXI) brought up the fact that the question was to secure a territory for the Assyrians. Some Assyro-Chaldeans continued to claim a protected zone in the plain of Nineveh (around Mosul) on their ancestral lands, which would be home not only to them but also to other non-Christian minorities, like the Yezidis.

After the creation of the Islamic State (IS) in 2006, but even before that, the question of a semi-autonomous territory belonging to the Christians and other minorities in northern Iraq was raised again: and it remains unresolved to this day. One cannot, however, seriously consider this question before being in a position to guarantee full security for the Assyro-Chaldeans, and, moreover, in the long term. In 2018 the Assyrian Confederation of Europe (ACE) opened an office in Brussels which deals with this subject. In 2016 the Iraqi parliament rejected the idea of revising the boundaries of the province of Nineveh (whose capital is Mosul) and of dividing it into several administrative entities.

We should remember that in 1925 Turkey had seized possession of the *vilayet* of Hakkari, and the *vilayet* of Mosul had been assigned to Iraq. In consequence, the territory in which many Assyrians had lived in the nineteenth century was divided between the present countries of Turkey and Iraq without taking into consideration the interests of the indigenous populations, including the Assyrian minority. As Claire Weibel Yacoub has written, the Assyrians' dream of self-determination and of reuniting again all their people on the same territory has been shattered. Hence the title of her book: *Le rêve brisé des Assyro-Chaldéens. L'introuvable autonomie* (The Shattered Dream of the Assyro-Chaldeans: an Unobtainable Autonomy). As so often in history, it was the interests of the great powers that had priority and imposed themselves. According to Henri Laurens, the stakes that the great European powers were playing for fashioned the political reality of the region and created terrible human tragedies, such as the quasi-destruction of Christianity in Turkey.

Concluding Remarks

In the course of the twentieth century the Assyrians were Christians who wandered from one region to another, caught in a vicious circle that drove them from what is today south-east Turkey (Hakkari), to Iran (Persia) and then to Iraq and to Khabur in the north-east of Syria. In 1933, leaving out of account the metropolitanate of Thrissur in India, one could scarcely still speak of a living Church of the East. The Assyrian people had been terribly decimated and the survivors found themselves without a territory. Moreover, the patriarch was sent into exile and subsequently lived outside the Middle East, far from his people's ancestral lands. It was thus that a Church in exile was formed.

Finally, according to the Assyrian patriarch Shimun XXI Eshai in a long message to his people on 20 November 1937, no less than three-fifths of the Assyrian people had been lost, for if one takes a rough starting figure of between 100,000 and 120,000 people, 60,000 to 72,000 had disappeared and so there were only 40,000 to 48,000 survivors. According to other statistics, if before the First World War the number of Assyrians was about 120,000, the war caused the disappearance of 40,000 people, that is to say, a third of the total pre-war population.

At Urmia the Assyrians were able to go home, but not in Hakkari, where they have never returned and have lost their ancestral lands. In Iraq the Assyrians lived in regions which the first Christians of their Church have inhabited since the beginning of Christianity in Mesopotamia. Those who escaped the massacres before, during and after the First World War, settled in Iraq and Syria as refugees. In Patriarch Eshai's own words, it seemed that the apostolic Church of the East was lost. As we have seen, some Assyrians decided to remain in Iraq and Persia (Iran); others managed to emigrate. This scattering allowed a new page to be turned in the history of the Church of the East: that of its interaction with the West and the rest of the world, with the organization of new parishes across the world in the diaspora.

CHAPTER 9

The Twenty-First Century around the World. The Diaspora

After the political events of the twentieth century that once again affected the history of the Church of the East so dramatically, the Assyrians in the Middle East were to be found principally in two countries, Iraq and Syria. From the 1980s the Christians, among them the Assyrians, started to leave Iraq as a result of a succession of wars: the Iran-Iraq war (1980–8), the Anfal campaign (1988) and the First Gulf War (1990–1). The pressures to emigrate have continued as a result of recent developments in the political situation of the Middle East, from the invasion of Iraq by the United States in 2003 and the attacks of Islamic State (IS) until the latter's fall, in Iraq in December 2017 and in Syria in March 2019.

Since the beginning of the twenty-first century the wars that have raged in the Middle East have affected Christians as a whole, whose numbers have been drastically reduced. In Iraq they were about 1.5 million (3% of the population) before the First Gulf War in 1991. Now their number is estimated to be less than 300,000. In Syria, before 2011, Christians represented at most from 8 to 10% of the population (about 1.6–2 million). There are no reliable statistics for 2019, but an optimistic estimate is that their present number does not exceed one million. This drastic demographic decline has of course affected the Christians of the Church of the East. These events as a whole explain the dispersion of a large number of Assyrians throughout the world, exiled and for the most part seeking refuge either in neighbouring countries (Jordan, Turkey and Lebanon) or elsewhere (Europe, North America and Australia) and seeking wherever possible to join their families already established abroad. Those who choose to stay in their land of origin are mostly adults, particularly the elderly.

This option is more difficult to imagine for the younger generations born in the Middle East or abroad.

In Iraq after the fall of Saddam Hussein in 2003, given their lack of security in Baghdad and elsewhere in the country (bombs in the churches, abductions, assassinations), many Christians, including Assyrians, sought refuge in the region of Mosul called the 'Plain of Nineveh' and in Iraqi Kurdistan (in northern Iraq). But in 2014 Islamic State (IS) captured the town of Mosul and its environs. It is thought that about 100,000 Christians at that time fled to Iraqi Kurdistan, which was close by, and also to neighbouring countries and elsewhere in the world. Furthermore, a unique cultural patrimony (churches and monasteries) was destroyed. In Syria a major armed conflict began in 2011 between different political and Islamic groups, one of which was IS, which made the town of Raqqa its capital (2014–17). Raqqa lies only 170 km from the village of Tall Tamer (along the River Khabur), the centre of the Assyrians of the Church of the East in north-east Syria.

Moreover, in the context of the diaspora, we may recall that Assyrians sought refuge in the Southern Caucasus (Armenia and Georgia) after the treaty of Turkmenchay (1828), which concluded the Russo-Persian war of 1826–8; some Assyrians settled there, while others emigrated to modern Russia and Ukraine. From there, a number emigrated to Europe and North America. During the Soviet Communist period (1917–91) their situation was difficult. Although some returned to Persia (called Iran after 1935), this was not the case for the Assyrians of Hakkari, who disappeared definitively from Turkey.

Outside the Middle East, diaspora communities of the Church of the East organized themselves in the course of the twentieth century on several continents, principally in North America but also in Europe, in the western countries of the ex-Soviet Union, and in Australia. In India, the ancient community there has never ceased to be active. Some of the faithful working in the United Arab Emirates and Qatar come under this Indian jurisdiction. At the beginning of the twenty-first century there were communities in all these continents in a number of countries.

With regard to Assyrian demographics in 1955, J. F. Coakley made the following estimates: 365,000 members of the Church of the East and 54,000 of the Ancient Church of the East (with its patriarchate at Baghdad under the authority of Patriarch Addai), making a total of about 419,000 Christians for the two Assyrian patriarchates. At the beginning of the twenty-first century, the Church of the East numbered about 400,000 members. To this figure should be added some 50,000 to 70,000 Christians belonging to the Ancient Church of the East, making a total of about 460,000 people.

What is the general organization of the Church of the East today? At a Holy Synod held on the 13 May 2019 at Aynkawa (close to Erbil) the following prelates were present, listed according to their respective countries: in Iraq, Patriarch Giwargis III and Bishop Abris Youkhanan of Erbil (for northern Iraq and also responsible for Russia and Armenia); for Iran, Bishop Narsai Benyamin; for Syria, Bishop Afram Athniel; for the United States of America, Bishops Awa Royel (of California), Paulus Benyamin (of the eastern United States), and Aprem Khamis (of the western United States); for Canada, Bishop Emmanuel Yosip; for Europe, Bishop Odisho Orahim (with his seat in Sweden); for Australia, New Zealand, and Lebanon, Metropolitan Meelis Zaya, accompanied by Bishop Benyamin Eliya (for the State of Victoria in Australia and for New Zealand); for India and the United Arab Emirates, Metropolitan Aprem (Mooken), with Bishops Youkhanan Yosip and Awgin Kuriakose. On the 26 May 2019 the first bishop of the Church of the East with his seat at London was ordained, Fr Narsai Youkhanis. It was announced that on the 3 November 2019 Fr Isaac Tamras would be ordained bishop of the diocese of Baghdad and would also be put in charge of the Assyrian communities in Ukraine and Georgia (<https://news.assyrianchurch.org/wp-content/uploads/2019/07/May-June-Voice-of-the-East-2019.pdf>). The recent history of the Assyrian communities in these countries up to 2019 is as follows.

Middle East

In Iraq the Assyrians have been indigenous to this geographical zone since the beginning of Christianity. In the twentieth century they inhabited the cities of Baghdad, Mosul and Kirkuk, or were dispersed among the towns and villages in the north of the country. This remained the case until the brutal events which at the beginning of the twenty-first century forced them to move elsewhere within Iraq or flee the country altogether. In 1981 Giwargis Sliwa (the present patriarch) was enthroned as metropolitan of Baghdad and all Iraq (under the name of Mar Giwargis). Since his ordination in 1999, Mar Isaac has been bishop for northern Iraq with his seat at Dohuk. At the very beginning of the twenty-first century the Assyrians lived in seven of Iraq's ten provinces: Dohuk (in the town itself and in the neighbouring villages), Erbil (mostly at Aynkawa, Diyana and Shaqlawa), Sulaymaniya, Nineveh, Kirkuk, Baghdad and Basra. In 2003 there were about 150,000 members of the Church of the East in Iraq, with four dioceses (Baghdad, Mosul, Kirkuk and Dohuk). In June 2014 IS captured Mosul, which was an important centre in the history of the Church of the East and the seat of its patriarch in the past. In August 2014 IS occupied the neighbouring plain of Nineveh, the last bastion of the Christians in Iraq. In 2015 the patriarchal seat of the Church of the East was moved from Chicago to Erbil, the capital of Iraqi Kurdistan. The cathedral of St John is situated in the town of Aynkawa (contiguous with Erbil), inhabited principally by Christians. The seat of the Assyrian patriarch of the Ancient Church of the East (the Old Calendarists) is at Baghdad. Since 1991 Iraqi Kurdistan has enjoyed a certain degree of autonomy with a regional government from 1992 at Erbil. We should remember that the town of Erbil was an important centre in the history of the Church of the East. In 2019 the largest Assyrian community was concentrated in the north of the country, in Iraqi Kurdistan, mostly at Dohuk and the surrounding district. At Dohuk, the Christian Aid Program in Northern Iraq (CAPNI) is the principal ecumenical organization that is helping the displaced populations in Iraq and Syria. Its director is the Assyrian priest, Fr Emmanuel Youkhana. Since 1972, cultural and linguistic rights have been accorded to Syriac Christians in Iraq. Some state schools and colleges have been established where courses are given in the modern Syriac language (Sureth) in towns where a certain number of Syriac Christians still live, particularly in Iraqi Kurdistan, and above all in the governorate of Dohuk (Duhok in Kurdish), where in 2015 there were thirty-eight schools with about 3,000 pupils. In 2006 a cultural association was founded with the aim of preserving the Syriac language, and Syriac history, and culture, called the Assyrian Writers League, which publishes, among other things the magazine *Ma'alta* (twice yearly in Syriac, Arabic and English). There are also radio and television programmes for Syriac Christians. In Iraqi Kurdistan two television channels broadcast in Sureth and also in Arabic and Kurdish. Each of these channels is affiliated to a political party: Ishtar TV to the Chaldean Syriac Assyrian Popular Council; and Ashur TV to the Assyrian Democratic Movement.

In Iran an episcopal see was established at Tehran in 1962, with Mar Dinkha, the future patriarch, at its head. In 1966 the cathedral of Mar Giwargis (St George) was consecrated and an Assyrian school was opened. The present bishop, Mar Narsai Benyamin, born in Urmia in 1980, was enthroned in 2010. Even before the Revolution of 1979, some Assyrians had left the Urmia region (the centre of the province of western Azerbaijan), where most of the Assyrians lived, to go and live in the capital Tehran and elsewhere for economic reasons. In 2019 there were 10,000 to 12,000 faithful in the whole country. There are three churches in Tehran: St George (the cathedral), St Mary and St Thomas. A priest resides in Tehran, another in Fardis (35 km from Tehran, in the province of Alborz), who officiates at a chapel constructed by the faithful (about 200 families), and two priests are responsible for Urmia and its environs. At Urmia and in the neighbouring villages there are about eighty churches belonging to the Church of the East. At Urmia itself, in the cathedral of St Mary, the Liturgy is celebrated every Sunday, and in the villages at least once a year on the feast day of the church's dedication (a saint or martyr). In the south, in the province of Ispahan, thirty or so families live at Shahinshar. Following the revolution of 1979, the Constitution of the Islamic Republic of Iran recognizes Christians of the Armenian and Assyrian traditions as religious minorities. A seat is reserved for the Assyrians and Chaldean Catholics in the parliament (*Majlis*), a seat occupied in 2019 by Yonathan Betkolia of the Church of the East, who has been re-elected since 2000, and who is also the secretary general of the Assyrian Universal Alliance (AUA). His grandfather, the priest Shmoeil Bektolia, who lived in the village of Gogtapa near Urmia, was a writer of repute. The Syriac language is taught at the Assyrian church and school of Tehran. Within the context of a heritage policy, the Iranian authorities consider certain ancient churches as historical monuments that should be preserved as part of the national heritage. The government has offered subsidies to help the Church of the East restore some of its churches near Urmia. As explained in the preceding chapters, the presence of the Church of the East in Iran (formerly Persia) is very ancient. In the churches and on the graves in the cemeteries there are an important number of more or less ancient Syriac inscriptions.

With regard to Syria, we may recall that a diocese of the Church of the East was established at Damascus in the seventh century and was then elevated to the rank of a metropolitanate under the patriarch Timothy I (780–823). At the end of the ninth century this metropolitanate included the following suffragan dioceses: Aleppo, Mabbug, Mopsuestia, Tarsus, Melitene (Malatya), and Jerusalem. After 1933 the Assyrians came to settle in north-eastern Syria along the river Khabur (to the west of the town of Hassake), where they built thirty or so villages. Some of them settled elsewhere, principally in the towns close to Hassake and Qamichli (on the Turkish frontier, near the ancient city of Nisibis, now called Nusaybin, in Turkey). In 1968 Mar Yohannan Abraham (d. 1985) was enthronized as bishop of Syria (which took place in the church in Beirut) with his seat at Hassake. The present bishop of Syria, still with his seat at Hassake, Mar Aprem Athaniel, was enthronized in 1999. In February 2015 Islamic State (IS) attacked the villages of Khabur and 228 Assyrians were kidnapped for ransom. On 23 September 2015, IS executed three of them and showed it on a video. About 5,000 Assyrians were obliged to leave their villages, which were pillaged, burnt and demolished, as were the churches. In 2019 only about 1,000 faithful remained out of 20,000 before the war. Some parishes are still active at Hassake, Qamichli and Tall Tamer, the largest village of the Khabur. In 2015 the Assyrian International News Agency (AINA) published a list of churches and monasteries destroyed by IS and other Muslim groups in Syria (<http://www.aina.org/news/20151212211531.htm>).

In Lebanon some Assyrian migrants from Iraq and Syria came to settle in the 1950s, principally in Beirut. In February 1962 the patriarch Shimun XXI Esai went to Beirut, where about 2,000 Assyrians lived. He briefly visited Jerusalem and then returned to Beirut to consecrate the church of St George, the present cathedral situated in the Boshrieh quarter of the city. In Beirut in 1968 the patriarch enthronized Mar Narsai of Baz (d.

2010) bishop of Lebanon. In 1976 Mar Narsai was promoted to the rank of metropolitan of Lebanon, Syria and Europe. After his death in 2010 he was replaced in Lebanon by Mar Meelis Zaya, the present metropolitan of Australia and New Zealand. In 2019 there were about 3,000 Assyrians in Lebanon, distributed among four parishes: three in Beirut (at Boshrieh, Ashrafieh and Hadat) and one in Zahle.

In Jordan an Assyrian priest from Iraq goes regularly to Amman to minister to the Assyrian refugees and preside at the Liturgy. There is no parish, properly speaking. The religious services are celebrated in the parishes of the Syrian Orthodox Church.

In Israel there was no community of the Church of the East in Jerusalem in 2019. In the sixteenth century the presence of the faithful of this Church is recorded at the Holy Sepulchre and in Palestine. There was apparently a church in Jerusalem in the eighteenth century.

North America

As already noted, under the patriarch Dinkha IV Chicago was the patriarchal seat from 1976 to 2015, the year when it was moved to Erbil. In North America Assyro-Chaldean refugees (together with some Assyrian Protestants) began to arrive in the United States at the end of the nineteenth century, but most of all in the aftermath of the First World War (1914–18). One can get an idea of the conditions of life of these first emigrants by consulting the weekly newspaper *Star* (*Kokhva*), which was published in Urmia between 1906 and 1918. In 1913 Fr Odisho arrived in the United States from Persia in 1915 and established a parish in Chicago. He then became the co-founder of the parishes of Gary (Indiana) in 1919, New Britain (Connecticut), Detroit and Flint (Michigan), and also Turlock and Fresno (California). In 1944, in a report on the distribution of the faithful in the United States, Patriarch Shimun gave the following estimates: Chicago, Illinois 5,000; Gary, Indiana, 1,000; Flint 600 and Detroit 100, Michigan; New Britain, Connecticut, 1,200; Elizabeth, New Jersey, 200; city of New York, New York and region, 500; Philadelphia,

Pennsylvania, 500; Turlock and San Francisco, California, 1,500; with a total of about 11,000 people.

In 1976 Mar Aprem Khamis, who was then bishop of Basra (Iraq), was appointed bishop responsible for the United States and Canada. In 1984 two bishops were enthroned for the United States by Patriarch Dinkha IV in the cathedral at Chicago: Bawai Soro was ordained for the diocese of the western United States, and Aprem Khamis for the eastern United States and Canada.

In 2019 some estimates of the number of members of the Church of the East in North America put the figure at 200,000 persons in the United States – the most numerous community in the diaspora. We should remember that it was after the forced exile of Patriarch Shimun that the patriarchal seat was transferred to Chicago, Illinois, in the United States. That is why a large number of Assyrians came to settle above all in Chicago, a great industrial city which along with its environs became the centre of gravity for the Assyro-American diaspora. Other Assyrians settled in the neighbouring state of Michigan, in the towns of Detroit and Flint. Assyrian communities are found in other states, above all in California and Arizona in the west, but there are also a number of faithful belonging to the see of Chicago on the east coast. In 2019 there were three bishops: Mar Paulus Benyamin, responsible for the east of the United States, with his seat at Chicago; Mar Awa Royel, secretary of the Holy Synod and responsible for California, whose seat is at Modesto (with six parishes in California and one in Seattle in the state of Washington); and Mar Aprem Khamis, responsible for the west of the United States, with his seat at Phoenix, Arizona (six established parishes and three parishes called 'mission parishes' in a state of development). In Chicago and its environs there were in 2019 ten parishes. In the Chicago region there are also a number of associations, among them the Assyrian American Association of Chicago, which was established in 1917. In Chicago is also the Assyrian Universal Alliance Foundation (AUAF) founded in Pau, France, in 1968. At the AUAF in Chicago, the Assyrian Heritage Museum can be visited that was founded by Helen Schwarten (d. 1999), and also the Ashurbanipal Library, which was opened in 1986 and houses an important collection of Assyrian texts. In Chicago one can also visit the celebrated Oriental Institute Museum, which houses a rich collection of archaeological objects from the Middle East, among them finds from ancient Assyria. On the east coast the main communities of the Church of the East are at Yonkers (New York) and New Britain (Connecticut). In California, at Turlock, the parish of Mar Addai was founded in 1950. In the neighbouring town of Modesto, the church of Mar Zaya was consecrated in 1990. At San Francisco the church of Mar Nasai was consecrated by Patriarch Shimun in 1958. In California other parishes were also organized. At Los Angeles a church was bought by the community in 1984. At San Diego the parish of St Hormizd in Spring Valley began its activities in 2011, and the church was consecrated in 2014. At Ceres the church is dedicated to St Addai. At Valencia a new parish was started in 2016. In Arizona the Assyrians settled in Phoenix. The cathedral of St Peter was consecrated in 2007 and is situated in the town of Glendale. At Gilbert in Arizona a second church was organized and dedicated to Mar Yosip (St Joseph) in 2019. At Las Vegas in Nevada the church of Mar Benyamin was established in 2010. In 2019 two parishes were in the course of organization at Houston and Dallas, Texas.

In the United States there are a number of clubs and organizations, and also magazines, as well as the radio and television channels intended for Syriac Christians. There are also several publications. The *Assyrian Star*, the official magazine of the Assyrian American National Federation, was first published in 1972. In 1948 publication began of *Light from the East*, a magazine disseminating Church news. The *Journal of Assyrian Academic Studies* (*JAAS*) is a university journal dedicated to Assyrian studies. The *Voice from the East*, a Church magazine, was published in Chicago from 1982 to 1995. In California the television channel TV 23 Assyria Vision began broadcasting on digital UHF channel 15 in 1996 in the region of Modesto, Stockton and Sacramento. Sometimes it discusses matters relating to the Church. The Assyrian Christians of the United States meet once a year for a large national Congress, the Assyrian American National Federation Convention. Among the Assyrian personalities of the United States, two may be mentioned here: John J. Nimrod (1922–2009), a senator of Illinois and activist for human rights, and the historian Eden Naby.

In the whole of Canada in 2019 there were about 25,000 people of the Church of the East. The largest number lives in the province of Ontario, especially in Toronto. In 1979 a small church was purchased in this city, served by a priest coming from the United States. It was sold and another church was bought in 1982 at Mississauga. This was the first official parish, but this church too was sold in 2004. Then the newly built cathedral of St Mary at Toronto opened in 2006. In 1986 a second parish was established in London, Ontario. In 2019 there were two other parishes in Ontario, at Hamilton and Windsor. There were also three 'mission parishes' at Calgary-Edmonton in Alberta, at Vancouver in British Columbia and at Kitchener in Ontario. The bishop responsible for Canada is Mar Emmanuel Yousip.

Europe

In Western Europe the history of the Church of the East began in England, in London, in the 1950s with Assyrian Christians coming principally from Iraq. In Sweden at the beginning of the 1970s the Assyrians came mostly from Syria; the first priest was ordained at Jönköping in 1974. In Germany the number of Assyrians coming from Iraq and Syria grew during the 1970s; in 2019 it was the largest community in Western Europe. In the south (in Greece and Italy) most of the Assyrians are refugees from the Middle East (from Irak and Syria) in transit to other countries. In 2019 in the whole of Europe there were about 20,000 Assyrians. The details are as follows.

In Great Britain the community is centred in Ealing in West London. There are two parishes in Ealing, one under the jurisdiction of the Church of the East, the other under the jurisdiction of the patriarch at Baghdad (the Old Calendarists), with about 500 families. In Germany the community (numbering 7,000 to 8,000 persons) is spread out throughout the country, with parishes at Wiesbaden, Munich and Borken, where a chapel was purchased and consecrated in 2014. In Sweden there are seven parishes, with six priests and a good number of the faithful but a little fewer than in Germany. In the following countries the faithful are less numerous and do not benefit from the presence of a local priest (a priest from a neighbouring country visits the community once or twice a month, or more rarely). These were the approximate numbers in 2019: in Finland 150; in Norway 500; in Denmark 500; in Austria 450 (with parishes in Vienna, Linz and Graz); in the Netherlands 400 (with two parishes, in Zeist and in Oldenzaal); in Belgium 100 families (with a parish in Brussels); in France 450 in Marseille; and in Greece more than 200 families. In 2018 the whole of Western Europe was placed under the jurisdiction of Bishop Odisho Orahim, whose seat has been in Sweden at Huddinge (near Stockholm) since 1994. During the Holy Synod of May 2019 it was decided that Western Europe should be divided into two dioceses, each with its own bishop: a diocese of Western Europe for the faithful of the United Kingdom, France, Belgium, Austria, the Netherlands and Greece, under Mar Awraham Youkhanis (originally of Australia), who was enthroned as a bishop in May 2019, with his seat in London; and another European diocese of Scandinavia and Germany, remaining under Mar Odisho Orahim, who became responsible for the faithful of Norway, Finland, Denmark and Germany, with his seat remaining in Sweden.

Ex-Soviet Union

During the Soviet period an atheist education was imposed on all. The Assyrian Christians too suffered religious persecution and some of them were deported to camps in Siberia and Kazakhstan. During the 1920s there were up to 50,000 Assyrians in the Soviet Union, some of whom later emigrated to Europe and North America. Almost two thirds chose to remain in the new Soviet states in Transcaucasia (according to some statistics, 2,904 in Georgia, 21,215 in Armenia and 964 in Azerbaijan). By 1959 their number had gone down to less than half. Since the 1990s (the years of *perestroika*), the situation has changed and the Church of the East has begun to organize itself. Despite a certain tendency to become immersed in Russian culture, the community has rediscovered its roots and is trying at the present time to preserve and develop its Assyrian identity.

In Russia, most of the Assyrians are in the south-west, in the Caucasus region. In 2019 some estimates of the population of Assyrian origin put the figure at more than 50,000 persons. Because for a long time they had neither priests nor parishes, a great number of them became members of the Russian Orthodox Church (patriarchate of Moscow), which organized parishes for them in Russia and other countries. The Church of the East has regular official contacts with the patriarchate of Moscow. The patriarch Dinkha IV paid his first visit to Russia and Ukraine in 1981, when he went to Moscow and Kiev. In 1994, while he was still metropolitan of Iraq, the present patriarch, Giwargis, made his first pastoral visit to Russia and also became the bishop responsible for the country. Since 1996 a permanent priest of Iraqi origin, Kamis Yosip, has been assigned to a new parish established in Moscow, where the construction of a church (Mart Mariam) was completed in 1998. In 2019 there were two parishes in Russia. In Moscow the church of Mart Mariam was consecrated in 1998 by Patriarch Dinkha IV. In Krasnodar (1,200 km south of Moscow) the church of St George (Giwargis) was built in 1998 and also consecrated in 2014 by Patriarch Mar Dinkha IV, who at the same time blessed the foundation stone of a future church at Rostov dedicated to St John the Baptist. Among erudite Assyrian personalities in Russia, Mikhail Sado (1934–2010) may be mentioned, professor of Semitic languages at St Petersburg. He spent thirteen years (1967–80) in prison during the Soviet era because of his religious activism. His son, Stephen Sado, became a monk and priest of the Russian Orthodox Church (patriarchate of Moscow).

The South Caucasus (or Transcaucasia) has always been a land of asylum for the Assyrians and other Eastern Christians. The Assyrians came there principally from Persia and the Ottoman Empire as a result of various waves of persecution (discussed in earlier chapters), but also for economic reasons. In fact after the Treaty of Turkmenchay (1828) thousands of Assyrians used to come to the Russian provinces of the Caucasus as seasonal workers.

In Armenia the Assyrians began to arrive in the 1830s and were concentrated in villages not far from the capital, Yerevan. In 2019 there were about 3,000 Assyrians. The majority of them live at Verin Dvin (37 km south of Yerevan) and also in the villages of Arzni (19 km north of Yerevan) and Nor Artagers (56 km west of Yervan, formerly called Shahriyar). Some live in Yerevan and elsewhere. The church of Verin Dvin (called Dvin-Aysor in Soviet times), established at the time of the arrival of the Assyrians in 1828 and dedicated to St Thomas, was restored in 2001. The church of Arzni was built in 1828. In 2019 a single priest, Fr Nikadim Yukhanaev, served these communities. Born and resident in Verin Dvin, he was ordained in 2014. Before him, Fr Isaac Tamraz had been sent from Iraq to serve the church from 2003 to 2013. In these villages where there is an Assyrian population the modern Assyrian language is taught at school at all grades. There are some Assyrians who became members of the Russian Orthodox Church and have their parish in the village of Koilasar (formerly Dimitrov), between Yerevan and Dvin.

In Georgia, the first community of the Church of the East was concentrated at the capital Tbilissi. Before the First World War (1914–18) and during the 1930s this was the most active Assyrian community in the Soviet Union, but by 2019 there was no longer any parish belonging to the Church of the East. There are about 2,000 people, most of whom live in the village of Dzveli Qanda (30 km to the west of Tbilissi, near the motorway going to Gori) and speak Syriac. There they pray in the church and in the adjacent monastery dedicated to the Thirteen Assyrian Saints. These saints are monks of the Assyrian tradition known to the Georgians for having founded the first monasteries in Georgia in the sixth century. In 2019 this place belonged to the Georgian Orthodox patriarchate and was directed by the founder, Archimandrite Seraphim Betgharibi. Although his mother was originally of the Church of the East, he in fact became a monk of the Georgian Orthodox Church. The Orthodox Liturgy of St John Chrysostom is used there, translated into Syriac.

In Ukraine there are no available statistics relating to the Assyrians. They live principally in the cities of Kiev and Lviv, and also in Donetsk, Dnipro, Kharkiv, Krasnoarmeisk, Zaporijia, Mariupol and Odessa. In 2019 they had neither a priest nor a parish. Nevertheless, some of the faithful meet occasionally to pray together. Until 2019, an Assyrian bishop or priest used to visit them from time to time. Most of the Assyrians have joined local Orthodox parishes.

Australia

In Australia, Assyrian migrants arrived mostly as refugees in successive waves from Iran, Iraq, Syria and Lebanon. This immigration began in the 1960s. The Assyrian communities settled above all in Sydney and Melbourne. There is a small number of families in Brisbane and Perth, without a parish. In 2019 there were about 23,000 members of the Church of the East in Sydney and 7,500 in the State of Victoria. The majority live to the west of Sydney in the suburbs of Fairfield (including Smithfield), Cecil Hills and Greenfield Park in the State of New South Wales. The first priest sent to Sydney from Lebanon in 1971 was Fr Giwargis Yonan. The first church in Australia was that of St Mary in Polding Street (Smithfield), consecrated in 1974. The parish of Melbourne was established in 1980. In 1984 the present archbishop, Mar Meelis Zaya, was ordained bishop for the diocese of Australia and New Zealand and arrived in Sydney in March 1985. In December 2008 he became metropolitan of Australia, New Zealand and Lebanon. In 2019 the parishes dependent on Mar Meelis were all situated in New South Wales, near Sydney, namely, the cathedral of St Hurmizd in Greenfield Park, consecrated in December 1990, and the churches of St Mary and Sts Peter and Paul in Cecil Park. The latter was consecrated in 2014. The liturgy is celebrated in English there and there are regular meetings of young people. In Australia the youth organizations are particularly well organized. Another parish was to open in 2019, that of Mar Yusip in Leppington. In Matraville a church has been leased for the parish of St George. Archbishop Zaya prioritizes education and the young, and also the care of the elderly. In 2012 a retirement village was opened near the church of St Mary in Smithfield with fifty-one homes. With regard to education, in 2002 Patriarch Dinkha IV inaugurated the first school (named St Hurmizd) belonging to the Church of the East outside the Middle East. This elementary school with 740 pupils developed into a secondary school in 2019. In 2006 the St Narsai Assyrian Christian College in Horsley Park was officially opened. On the same site the Nisibis Assyrian Theological College began offering courses in 2020, with accreditation from Sydney College of Divinity.

Other projects are in the pipeline. In 2016 the Assyrian Language College was opened, where hundreds of young people and adults follow courses on their maternal Syriac language.

In Melbourne the earliest parish is located in Reservoir. Dedicated to St George, it was founded in 1979 and its first priest was the late Fr Younan Towana. In the northern suburb of Coolaroo the cathedral of St Abdisho was consecrated in 2014. In 2017 a bishop was enthroned, Mar Benyamin Elya, for the new diocese of Victoria and New Zealand, whose seat is at Melbourne. In New Zealand Mar Meelis Zaya organized the first community in 1986. The faithful are originally from Syria, Lebanon, Iran and Iraq. In 2019 there were 3,000 Assyrians in New Zealand, with two parishes, St Mary in Manurewa (in South Aukland), which was the first to be opened (in 2014), and St Odisho in Wellington. The first priest, in 1992, was Fr Toma in Aukland.

In Australia there are also parishes belonging to the jurisdiction of the Ancient Church of the East, under the patriarch of Baghdad, which are supervised by Archbishop Yaqob Daniel. The faithful have built their cathedral dedicated to St Zaya in Middleton Grange, a suburb of Sydney. They also have a church in Melbourne and another in Wellington (New Zealand).

India

The community of the Church of the East has been present in India since the first centuries of Christianity. This Church was reorganized in 1908 by Metropolitan Abimelek Timothy (d. 1945). In 2019 it numbered about 25,000 members. Most of them live in Kerala, mainly in Thrissur (the city which is the seat of their metropolitan) and in the neighbouring large city of Kochi. Others live in the great Indian metropolises, such as Delhi and Coimbatore, and also in Mumbai (formerly Bombay), Chennai (formerly Madras) and Bangalore, where the faithful have built their own churches. In all, in 2019 there were thirty-two parishes in India, twenty-six of which are in Kerala, with forty-nine priests. The present metropolitan of India is Mar Aprem (Mooken). In 2019 he was assisted by two auxiliary bishops,

Mar Awgin Kuriakose and Mar Youkhannan Yosip, the latter being responsible for the theological seminary at Thrissur, constructed in 1956 by Metropolitan Thoma at Mulangunnathukavu (10 km from Thrissur). In 2019 three nuns lived at Thrissur in a small monastery founded in 1998, working with orphans. There were two orphanages (the first founded in 1962), two homes for the elderly, and also a hospital. The Church in addition administered four schools and two colleges. An important printing press has been active since 1926. The Liturgy is celebrated principally in the local language, Malayalam, which is understood by all the faithful, while keeping a small part in Syriac. Metropolitan Mar Aprem has written a small book, *Teach Yourself Aramaic* (1981), and also a number of other works. The archdiocese of India publishes the magazine *Voice of the East* (<https://news.assyrianchurch.org/voice-of-the-east/>).

In the United Arab Emirates Mar Aprem inaugurated a parish in 2005 at Sharjah for the faithful from India, but also from Iraq, Iran, Syria and Lebanon, working in this region. The Liturgy is celebrated each Friday in the St Philip chapel of the Anglican church of St Martin. Liturgies are celebrated from time to time at Muscat (Oman) and Doha (Qatar). In 2019 the bishop responsible for this region of the Arab Emirates and Muscat (Oman) was Mar Yokhannan Yosip.

In conclusion, as we have seen, the exodus of Assyrian Christians has continued in the twenty-first century right across the world. As a result of this, an Assyrian population has developed in all the continents, above all in North America, Europe and Australia. All of them have sought a peaceful life in diaspora. These Assyrians were forced to go into exile, probably, for the most part, without hope of return to the Middle East, to the lands of their illustrious ancestors, at least until conditions of security and freedom can be stabilized for the long term.

Conclusion

As we have seen in this book, from the geographical point of view the history of the Church of the East until the nineteenth century is an *Asian* history of an *Asian* Church. From the beginning from the twentieth century, this history, still in geographical terms, has expanded globally, around the world. It is worth remembering that Jerusalem, Antioch and many other places in the Middle East cited in the Gospels, and subsequently in the history of Christianity, are in Western Asia. We should also bear in mind that the notions of east and west are relative to a given geographical situation: one is always to the east or to the west in relation to someone else. As the British historian Peter Frankopan explains in his book *The Silk Roads: A New History of the World*, this relativity obliges us in the West to abandon a Eurocentric vision and understand that the centre of the world may be 'elsewhere'. Our study adopts the same perspective with regard to the history of the Church of the East. If, for Frankopan, the history of Asia is central to the history of the world, one may equally consider the history of the Church of the East important in the history of the universal Church, for it concerns a very ancient Christian community whose history is unique in the history of Christianity and very little discussed. Today it is still an international Church, scattered around the world. In fact the great majority of the faithful live in the diaspora and not in their ancestral lands of origin in Mesopotamia. In its active missionary phase, when the faithful belonged to a large number of ethnic groups and spoke a variety of languages, this Church had already put into effect what we now call 'inculturation', carrying out, among other tasks, the translation of texts into different languages for the peoples to whom it had brought the Gospel.

In this book I have attempted to explain, somewhat sketchily, the history of the Church of the East and its communities down to our own day. During the centuries, repeated episodes of massacre and forced exodus have led to the settlement of these Christians since the fourteenth century principally in northern Iraq and south-east Turkey. In our own time, the violence that has continued in the Middle East explains why the faithful, members of the clergy and prelates of the Church of the East have been forced to go into exile and form a diaspora. Among the difficulties encountered by the Church of the East are those connected with external factors, such as the persecutions it experienced, and also internal conflicts, both leading to its weakening. In the twentieth century it was the game played by the great European powers that destabilized the Middle East. The Assyrian Christians were among the victims.

After the decline of the Church of the East since the fourteenth century, this Christian community became a religious minority which had to struggle very hard to survive. Constant moves, the loss of their land and their property, insecurity, violations of their human rights and religious freedom, martyrdom and genocide and also the persecutions in the nineteenth century and until our own day – the faithful of the Church of the East have been, and still are, victims of all this. Some Assyrian Christians say that they lost everything except their life and their faith. At the end of the First World War (1918) and on the disintegration of the Ottoman Empire (1923), the protection of minorities and the fate of the Assyro-Chaldean people, became one of the most complicated problems to deal with and remains unresolved.

Among the Assyro-Chaldeans there emerged, towards the end of the nineteenth century, the consciousness of constituting a 'nation'. A certain Assyrian nationalism developed in response to a search for identity, sometimes to the point of claiming an affiliation with the ancient Assyrians. Since the genocide of 1915, the 'Assyro-Chaldean question' has only become more complex. The indigenous Assyro-Chaldeans have been driven out of their ancestral territories and many have become refugees throughout the world. This phenomenon of exile persists in the twenty-first century, above all as a result of the recent wars waged successively in the Middle East, particularly in Iraq and Syria. There are however, many who today stress the importance of the Christians of the East remaining in their countries of origin, whenever that is possible.

Conclusion

What about the future? At the present time, for Eastern Christians to be offered a future in their countries of origin in the Middle East, a stable peace needs be established in the region, permitting them an existence with some security by the imposition of fundamental concepts of religious freedom, civil equality and human rights. These Christians need to be supported in different ways, which is already being done by certain NGOs and associations.

With regard to the diaspora, their struggle is not just for physical survival but also for the preservation of their identity and culture. On this subject Fr Emmanuel Youkhana emphasizes that one cannot hope to preserve one's identity without reference to one's country of origin and without the mother-Church where one's roots are. That is why it is essential that the Assyrian community should struggle to continue to exist in its own native country (in the Middle East), and that the link between the diaspora and the country of origin should always be maintained. Besides, how can one forget one's country of origin? As an Assyrian of Syria said, 'Do not leave, but if you do leave, try to return!' This is what makes some Assyrians of the diaspora visit their ancestral villages and churches during their vacations, going as tourists and combining this with making pilgrimages and returning to their roots. Even if these communities in the Middle East and the diaspora are separated from each other by great distances, modern means of communication, particularly through the internet, allow them to maintain links between them and a feeling of common belonging.

In the diaspora the Assyrian people are faced with other difficulties. Notably, they do not escape the so-called 'melting-pot' phenomenon, that is to say, the ethnic, religious and cultural standardization resulting from a policy of assimilating numerous immigrants of different origin in our globalized societies. In these conditions it is not easy to keep one's identity and culture, the two being inseparable. How do you preserve them in any case when the fact of immigration already constitutes a loss of evident identity? In 1921 Basile Nikitine wrote that we should admire the astonishing vitality which has allowed the faithful of the Church of the East to keep their language, their religion, and their national physiognomy intact throughout the centuries. And Georges Bohas wrote in 1994 that this identity has persisted above all in language and religion. The Church thus plays a very large role in preserving all this. Indeed, the bishops of the Church of the East have already, at the end of the twentieth and the beginning of the twenty-first centuries, organized parishes and centres in accordance with what is needed in the Middle East and the diaspora.

In the diaspora Assyrian Christians are certainly influenced by the way of life of their country of adoption, all the more so because they are faced with the necessity of adapting themselves and integrating in the best and quickest way possible. Their new social relations and, of course, mixed marriages contribute further to the weakening of their cultural particularity. It is ultimately in the context of family and parish life, through the influence of parents, grandparents and clergy, that Assyrian culture and identity are best transmitted.

Identity is linked to history and vice versa. Dispersed for many years to the four corners of the earth, how can the Christians of the Church of the East succeed in preserving their language, their culture and their faith in the long term? The classical Syriac language, too, is profoundly a part of the identity of the faithful of the Church of the East. Indeed, in order not to forget one's roots and to understand them more profoundly, one must not forget the language. And even if some families still speak the vernacular Syriac language, Sureth, its existence seems threatened in the long term. The Church of the East and some teachers make efforts to maintain the two languages (spoken and ancient) by giving courses in Sureth and Syriac, notably in the parishes, schools and some associations. As regards the transmission of a living culture, social events such as weddings and Assyrian festivals celebrated within families and also in the Assyrian clubs and associations, keep alive the traditional songs and dances with their colourful costumes, along with the culinary traditions.

What can be done to ensure that the Church of the East remains the guardian of the rich cultural, liturgical and spiritual heritage which belongs to it? It is for all its members as a body to respond to this question. Moreover, as the archbishop of India, Mar Aprem, told me: 'We must not hide our great tradition, but we should share it with others.' It is also for those who are not Assyrians to encourage this community to keep its roots alive in different ways.

Conclusion

The hope of the Christians of the Church of the East rests above all in the new generations. It is for this reason that I dedicate this book to them, that they may invest as much as possible in their Church and their communities. They must be encouraged to study the history of their ancestors and their geographical roots in depth, so as to keep alive the heritage that they have received: for example, by deepening their study of the language in both its ancient and modern forms (Syriac and Sureth), by pursuing historical and archaeological research at university level, and by preserving oral history through recording the memories of the witnesses of recent historical events. This implies putting the emphasis on teaching and on developing cultural programmes.

With regard to archaeology, the discoveries in this domain are without doubt constantly teaching us more and more about the richness of the history and patrimony of the Church of the East in Asia. Although a number of sites have been destroyed, others still remain to be discovered. What is at stake is to preserve the heritage of the Church of the East in the greatest degree possible, on the levels of both its material and non-material patrimony. It needs to be well understood that in the Middle East there have sometimes been attempts to obliterate the history of the Church of the East, not only by killing and driving out the faithful and their clergy, but also even by destroying the physical patrimony (churches, monasteries, and villages). In south-east Turkey and elsewhere only some ruins and some names witness to this past. This programme of cultural destruction is pursued, for example, through the modification of all the names of historical Christian villages. The Assyro-Chaldean past is thus eliminated even on the maps of some Middle Eastern countries, such as Turkey and Iraq, including Iraqi Kurdistan. It is for this reason that some speak of a patrimonial and cultural genocide.

Today those charged with the pastoral care of the Church of the East work ardently for the future of their communities in the Middle East and elsewhere in the world. At a meeting in September 2015 at Erbil after his enthronization, the catholicos-patriarch Giwargis told me during an interview: 'We shall do our best for the faithful of our church remaining in Iraq and the Middle East. I shall never lose hope and we shall do everything we possibly can to help and serve our people. We must focus on the spirituality of our ancient fathers of the Church, on transmitting it to our faithful and putting it into practice today. We must preserve the identity of our Church and our people by teaching them their language, their ecclesial tradition, and their culture.' A profound experience of spiritual life continues to be transmitted through reading the writings of the many Syriac fathers, the most celebrated of whom is Isaac of Nineveh whose universal spirit has moved all his readers down to our own day.

The transmission of identity means the preservation of an ethnic, linguistic, religious and even artistic heritage in the countries of origin and in the host countries of the diaspora. A loss of patrimony always generates a loss of identity. It is essential to guard one's identity and one's Syriac Christian patrimony and to transmit it to new generations. To love and transmit one's patrimony is an integral part of one's identity. One keeps a tradition, a faith and a culture alive by transmission. The Church of the East is there to help keep alive this transmission of a patrimony of twenty centuries.

Bibliography

Abbreviations

A Concise History = W. Baum and D. Winkler, *The Church of the East: A Concise History* (New York and London: Routledge, 2003).
Baumer 2016 = C. Baumer, *The Church of the East: An Illustrated History of Assyrian Christianity*, trans. M. G. Henry (London: I. B. Tauris, 2016).
BJRL = *Bulletin of the John Rylands Library*.
Briquel 2004 = F. Briquel Chatonnet, M. Debié, and A. Desreumaux (eds), *Les inscriptions syriaques* (Paris: Geuthner, 2004).
Briquel 2013 = F. Briquel Chatonnet (ed.), *Les églises en monde syriaque*. Études syriaques 10 (Paris: Geuthner, 2013).
Christian Heritage = E. Hunter (ed.), *The Christian Heritage of Iraq: Collected Papers from the Christianity of Iraq I-V Seminar Days*. Gorgias Eastern Christian Studies 13 (Piscataway, NJ: Gorgias Press, 2009).
CSCO = Corpus Scriptorum Christianorum Orientalium.
Études syriaques 12 = G. Borbone and P. Marsone (eds), *Le christianisme syriaque en Asie centrale et en Chine*. Études syriaques 12 (Paris: Geuthner, 2015).
From the Oxus River = L. Tang and D. Winkler (eds), *From the Oxus River to the Chinese Shores. Studies on East Syriac Christianity in China and Central Asia* (Berlin: LIT Verlag, 2013).
Gorgias Dictionary = S. Brock, et al., *Gorgias Encyclopedic Dictionary of the Syriac Heritage* (Piscataway, NJ: Gorgias Press, 2011).
Hidden Pearl = S. Brock (ed.), *The Hidden Pearl: The Syrian Orthodox Church and Its Ancient Aramaic Heritage* (Rome: Trans World Film Italia, 2001).
Hidden Treasures = D. Winkler and L. Tang (eds), *Hidden Treasures and Intercultural Encounters. Studies on East Syriac Christianity in China and Central Asia* (Münster: LIT Verlag, 2009).
JAAS = *Journal of Assyrian Academic Studies*.
Jingjiao = R. Malek and P. Hofrichter (eds), *Jingjiao: The Church of the East in China and Central Asia* (Sankt-Augustin: Institut Monumenta Serica, 2006).
ROC = *Revue de l'Orient chrétien*.
Syriac World = D. King (ed.), *The Syriac World* (London: Routledge, 2018).
Winds of Jingjiao = L. Tang and D. Winkler (eds), *Winds of Jingjiao. Studies on East Syriac Christianity in China and Central Asia* (Vienna: LIT Verlag, 2016).

Chapter 1. A History of the Church of the East: Origins to the Eighteenth Century

Baum, W. and D. Winkler, *The Church of the East: A Concise History* (New York and London: Routledge, 2003). <http://www.learnassyrian.com/assyrianlibrary/assyrianbooks/Religion/The%20Church%20of%20the%20East%20-%20A%20concise%20history%20-%20Wilhelm%20Baum%20and%20Dietmar%20W.%20Winkler.pdf>
Baumer, C., *The Church of the East: An Illustrated History of Assyrian Christianity*, trans. M. G. Henry (London: I. B. Tauris, 2016).
Bearman, P. et al., *Encyclopaedia of Islam* (Leiden: Brill, 1960–2008).
Bedjan, P., *Thomas of Marga's Monastic History and Other Texts* (Paris, 1901; repr. Piscataway, NJ: Gorgias Press, 2010).
Benjamin, D. D., *The Patriarchs of the Church of the East* (Piscataway, NJ: Gorgias Press, 2008).
Bertaina, D., *Christian and Muslim Dialogues* (Piscataway, NJ: Gorgias Press, 2011).
Borbone, P. G., 'Les "provinces de l'extérieur" vues par l'Église-mère', Études syriaques 12, 121–60.
Bosworth, C., *The New Islamic Dynasties: A Chronological and Genealogical Manual* (Edinburgh: Edinburgh University Press, 2014).
Brelaud, S., 'Un programme de restauration exceptionnelle au nord de l'Irak: les églises du monastère de Rabban Hormizd', in Briquel 2013, 381–90.
Brière, M., 'Histoire du couvent de Rabban Hormizd de 1808 à 1832', *Revue de l'Orient chrétien* 15:5 (1910), 410–24 and 16:6 (1911), 113–27.
Briquel Chatonnet, F., 'Writing Syriac. Manuscripts and Inscriptions', in *Syriac World*, 243–65.
Briquel Chatonnet, F. and M. Debié, *Le monde syriaque. Sur les routes d'un christianisme ignoré* (Paris: Geuthner, 2017).
Briquel Chatonnet, F., M. Debié, and A. Desreumaux (eds), *Les inscriptions syriaques* (Paris: Geuthner, 2004).
Brock, S., 'Isaac', *Encyclopaedia Iranica*, vol. 13, fasc. 6, 610–11. <http://www.iranicaonline.org/articles/isaac-bishop-of-seleucia-ctesiphon>
Brock, S., 'Christians in the Sasanian Empire: A Case of Divided Loyalties', in S. Brock, *Syriac Perspectives on Late Antiquity* (London: Variorum Reprints, 1984, chap. VI).
Brock, S., 'Du grec en syriaque. L'art de la traduction chez les Syriaques', in *Patrimoine Syriaque, Actes du Colloque IX. Les Syriaques transmetteurs de civilisation* (Antelias: Ordre, Antoine Maronite, 2005), 11–35.
Brock, S., 'A Guide to the Persian Martyr Acts', in S. Brock (ed.), *The History of the Holy Martyr Mar Ma'in* (Piscataway, NJ: Gorgias Press, 2008), 77–125.
Brock, S. et al. (eds), *The Hidden Pearl: The Syrian Orthodox Church and Its Ancient Aramaic Heritage*, 3 vols (Rome: Trans World Film Italia, 2001). Vol. 2, *The Heirs of the Ancient Aramaic Heritage*; see especially ch. 5, 'The Syriac Christian Tradition', 103–66; and ch. 6, 'The Spread of Syriac Christianity', 167–208.
Brock, S. et al., *Gorgias Encyclopedic Dictionary of the Syriac Heritage* (Piscataway: Gorgias Press, 2011).

Budge, E. A. W. (trans.), *The Book of Governors. The Historica Monastica of Thomas, Bishop of Marga* (London: Kegan Paul & Co., 1893; repr. Piscataway, NJ: Gorgias Press, 2003).

Budge, E. A. W., *The Histories of Rabban Hormizd the Persian and Rabban Bar-'idta*. Luzac's Semitic Text and Translation Series, vols 9–11 (London, 1902).

Bundy, D., 'Timotheos I', in *Gorgias Dictionary*, <https://gedsh.bethmardutho.org/Timotheos-I>.

Burkitt, F., *The Early Bishops of Edessa* (Piscataway, NJ: Gorgias Press, 2009).

Cabrol, C., 'Les secrétaires nestoriens à Bagdad (762–1258)', *Corpus des recherches arabes chrétiennes* 1 (Beirut, 2012).

Carlson, T., *Christianity in Fifteenth-Century Iraq* (Cambridge and New York: Cambridge University Press, 2018).

Carlson, T., 'Syriac in a Diverse Middle East: From the Mongol Ilkhanate to Ottoman Dominance, 1286–1517', in *Syriac World*, 718–30.

Cassis, M. C., 'Seleucia Ctesiphon', in *Gorgias Dictionary*, 365.

Chabot, J.-B., *Synodicon orientale ou Recueil de Synodes nestoriens* (Paris: Imprimerie Nationale, 1902).

Chaumont, M.-L., *La christianisation de l'Empire iranien : des origines aux grandes persécutions du IVe siècle* (Louvain: Peeters, 1988).

Coakley J. F., 'The Patriarchal List of the Church of the East', in G. J. Reinik and A. C. Klugkist (eds), *Orientalia Lovaniensia Analecta* 89 (Louvain: Peeters, 1999).

Daryaee, T., 'The Sasanian Empire', in *Syriac World*, 33–46.

Dauvillier, J., 'Les Provinces chaldéennes "de l'extérieur" au Moyen-Âge', in *Mélanges offerts au R. P. Ferdinand Cavallera* (Toulouse: Bibliothèque de l'Institut catholique, 1948), 261–316.

Debié, M., 'The Eastern Provinces of the Roman Empire in Late Antiquity', in *Syriac World*, 11–32.

Déroche, V., 'La fouille de Bazyan (Kurdistan irakien): un monastère nestorien ?', in *Briquel* (2013), 363–80.

Desreumaux, A., *Histoire du roi Abgar et de Jésus*, Apocryphes 3 (Turnhout: Brepols, 1993).

Desreumaux, A., 'Doctrine de l'apôtre Addaï', in F. Bovon and P. Geoltrain (eds), *Écrits apocryphes chrétiens* (Paris: Gallimand, 1997), 1473–525.

Dickens, M., *Echoes of a Forgotten Presence: Reconstructing the History of the Church of the East in Central Asia* (Münster: LIT Verlag, 2020).

Duval, R., *Histoire politique, religieuse et littéraire d'Édesse jusqu'à la Première Croisade* (Paris: Imprimerie Nationale, 1892; repr. Piscataway, NJ: Gorgias Press, 2012).

Erhart, V., 'The Church of the East during the Period of the Four Rightly-Guided Caliphs', *Bulletin of the John Rylands Library* 78:3 (1996), 55–71.

Fiey, J. M., *Mossoul chrétienne* (Beirut: Imprimerie Catholique, 1959).

Fiey, J. M., 'Balad et le Béth Arabayé irakien', *L'Orient Syrien* 9 (1964), 189–232.

Fiey, J. M., 'Assyriens ou Araméens ?', *L'Orient Syrien* 10:2 (1965), 142–60.

Fiey, J. M., *Assyrie chrétienne. Contribution à l'étude de l'histoire et de la géographie ecclésiastiques et monastiques du nord de l'Iraq*, vols 1 and 2 (Beirut: Dar El-Machreq Éditeurs, 1965), vol. 3 (Beirut: Dar El-Machreq Éditeurs, 1968).

Fiey, J. M., 'L'Élam, la première des métropoles ecclésiastiques syriennes orientales', *Paroles de l'Orient* 1:1 (1970), 123–153.

Fiey, J. M., *Jalons pour une histoire de l'Église en Iraq*, CSCO 310 Subs. 36 (Louvain: Peeters, 1970).

Fiey, J. M., 'Chrétientés syriaques du Horasan et du Ségestan', *Le Muséon* 86 (1973), 75–104.

Fiey, J. M., 'Adarbaygan chrétien', *Le Muséon* 86 (1973), 397–435.

Fiey, J. M., *Nisibe, métropole syriaque orientale et ses suffragants des origines à nos jours*. CSCO 388/54 (Louvain: Peeters, 1977).

Fiey, J. M., *Communautés syriaques en Iran et en Irak des origines à 1552* (London: Variorum Reprints, 1979).

Fiey, J. M., *Chrétiens syriaques sous les Abbassides surtout à Bagdad (749–1258)*. CSCO 420/59 (Louvain: Peeters, 1980).

Fiey, J. M., 'Résidences et sépultures de patriarches syriaques-orientaux', *Le Muséon* 98 (1985), 149–68.

Fiey, J. M., *Pour un Oriens christianus novus: répertoire des diocèses syriaques orientaux et occidentaux* (Beirut: Orient Institut Beirut, 1993).

Fiey, J. M., 'Isaac (catholicos)', in *Dictionnaire d'histoire et de géographie ecclésiastique*, vol. 26 (1997), 90–2.

Gignoux, P., 'La transmission de l'héritage grec aux Arabes par les Syriaques', in *Patrimoine Syriaque, Actes du Colloque IX. Les Syriaques transmetteurs de civilisation* (Antelias: Ordre, Antoine Maronite, 2005), 53–65.

Gillman, I. and H. J. Klimkeit (eds), *Christians in Asia before 1500* (Ann Arbor, MI: University of Michigan Press, 1999).

Harrak, A. (ed.), *The Acts of Mar Mari the Apostle*. Writings from the Greco-Roman World, 11 (Atlanta, GA: Society of Biblical Literature, 2005).

Hayes, E., *L'École d'Édesse* (Paris: Les Presses modernes, 1930; repr. Piscataway, NJ: Gorgias Press, 2012).

Herman, G., *Jews, Christians and Zoroastrians. Religious Dynamics in a Sasanian Context* (Piscataway, NJ: Gorgias Press, 2014).

Herman, G., 'The Syriac World in the Persian Empire', in *Syriac World* (Piscataway, NJ: Gorgias Press, 2014), 134–43.

Hunter, E. (ed.), *The Christian Heritage of Iraq: Collected Papers from the Christianity of Iraq I–V Seminar Days*. Gorgias Eastern Christian Studies 13 (Piscataway, NJ: Gorgias Press, 2009).

Jullien, C. and F. Jullien, *Les Actes de Mar Mari, l'apôtre de la Mésopotamie*. Apocryphes 11 (Turnhout: Brepols, 2001).

Jullien, C. and F. Jullien, *Apôtres des Confins. Processus missionnaires chrétiens dans l'empire iranien* (Louvain: Peeters, 2002).

Jullien, C. and F. Jullien (eds), *Les Actes de Mar Mari*. CSCO, 602/603 (Louvain: Peeters, 2003).

Jullien, C. and F. Jullien, *Aux origines de l'Église de Perse: Les Actes de Mar Mari*. CSCO, 604/114 (Louvain: Peeters, 2003).

Jullien, F., 'Édesse', in J. Leclant (ed.). *Dictionnaire de l'Antiquité* (Paris, 2005), cols 747–8.

Jullien, F., *Le monachisme en Perse. La réforme d'Abraham le Grand, père des moines de l'Orient*. CSCO 622 (Louvain: Peeters, 2008).

Jullien, F. 'The Great Monastery at Mount Izla and the Defence of the East-Syrian Identity', in E. Hunter (ed.), The Christian Heritage of Iraq (Piscataway: Gorgias Press, 2009), 54–63.
Jullien, F. (ed.) *Le monachisme syriaque* (Paris : Geuthner, 2010).
Jullien, F., *Histoire de Mar Abba, Catholicos de l'Orient*. CSCO 658 (Louvain: Peeters, 2015).
Jullien, F., 'Religious Life and Syriac Monasticism', in *Syriac World*, 88–104.
Kawerau, P., *Die Chronik von Arbela*. CSCO 199–200 (Louvain: Peeters, 1985).
Kessel, G., 'Syriac Medicine', in *Syriac World*, 438–59.
King, D. (ed.), *The Syriac World* (London: Routledge, 2018).
Labourt, J., *Christianisme dans l'empire perse sous la dynastie sassanide (224–632)* (Paris: V. Lecoffre, 1904).
Lamsa, G. and W. Chauncey Emhardt, *The Oldest Christian People: History and Traditions of the Assyrian People and the Fateful History of the Nestorian Church* (London: Macmillan, 1926; repr. Forgotten Books, 2013).
Landron, B., *Chrétiens et Musulmans en Irak: Attitudes nestoriennes vis-à-vis de l'islam* (Paris: Cariscript, 1994).
Leclercq, H., 'Perse', *Dictionnaire d'archéologie chrétienne et de liturgie*, vol. 14, cols 505–22.
Le Coz, R., *L'Église d'Orient: chrétiens d'Irak, d'Iran et de Turquie* (Paris: Éd. du Cerf, 1995).
Le Coz, R., *Les médecins nestoriens au Moyen-Âge: les maîtres des Arabes* (Paris: Harmattan, 2004).
Martin, J. P., *Les origines de l'Église d'Édesse et des Églises syriennes* (Paris: Maisonneuve et C. Leclerc, 1889; repr. Piscataway, NJ: Gorgias Press, 2011).
Meinardus, O., 'The Nestorians in Egypt: A Note on the Nestorians in Jerusalem', *Oriens Christianus* 51 (1967), 112–29.
Mellon Saint-Laurent, J. N., *Missionary Stories and the Formation of the Syriac Churches* (Oakland: University of California Press, 2015).
Moffett, S., *A History of Christianity in Asia*, vol. 1, *Beginnings to 1500* (San Francisco, 1992).
Morony, M., *Christians in Iraq after the Muslim Conquest* (Piscataway, NJ: Gorgias Press, 2006).
Murre-van den Berg, H., 'The Patriarchs of the Church of the East from the Fifteenth to Eighteenth Centuries', *Hugoye. Journal of Syriac Studies* 2:2 (1999), 235–64.
Murre-van den Berg, H., 'Hormizd, Monastery of Rabban', in *Gorgias Dictionary*, 203–5.
Nikitine, B., 'Nestoriens', *Encyclopédie de l'Islam* 3 (1936), 965–8.
Parry, K., 'Byzantine-Rite Christians (Melkites) in Central Asia and China and their Contacts with the Church of the East', in *Winds of Jingjiao*, 203–20.
Penn, M., 'Early Syriac Reactions to the Rise of Islam', in *Syriac World*, 175–88.
Platt, A., 'Changing Mission at Home and Abroad: Catholicos Timothy I and the Church of the East in the Early Abbasid Period', in *Winds of Jingjiao*, 161–82.
Potts, D. T., 'The Sasanians and Arabia', <http://www.iranicaonline.org/articles/arabia-ii-sasanians-and-arabia> (2012).
Reuther, O., 'The German Excavations at Ctesiphon', *Antiquity* 3 (1929), 449–50.
Scher, A., *The Famous School of Nisibis: A Brief History on Its Origins, Principals, and Its Most Prominent Teachers* (repr. Piscataway, NJ: Gorgias Press, 2011).
Segal, J. B., *Edessa, 'The Blessed City'* (Oxford: Clarendon Press, 1970; repr. Piscataway, NJ: Gorgias Press, 2005).

Segal, J. B., 'Abgar', *Encyclopædia Iranica*, I:2, 210–13. <http://www.iranicaonline.org/articles/abgar-dynasty-of-edessa-2nd-century-bc-to-3rd-century-ad>
Sellwood, D., 'Adiabene', *Encyclopædia Iranica*, I:5, 456–9 <http://www.iranicaonline.org/articles/adiabene>
Taylor, D., 'The Coming of Christianity to Mesopotamia', in *Syriac World*, 68–87.
Teule, H., *Les Assyro-chaldéens. Chrétiens d'Irak, d'Iran et de Turquie* (Turnhout, 2008).
Tisserand, E., 'Église nestorienne', *Dictionnaire de théologie catholique* 11:1 (1931), cols 157–218.
Tisserand, E., 'Église nestorienne, Listes patriarcales', *Dictionnaire de théologie catholique* 11:1 (1931), cols 259–63.
Tixeront, L. J., *Les origines de l'Église d'Édesse et la Légende d'Abgar* (Paris: Maisonneuve et C. Leclerc, 1888; repr. Piscataway, NJ: Gorgias Press, 2011).
Tubach, J., 'The Mission Field of the Apostle Thomas', in *Hidden Treasures*, 291–303.
Vaschalde, A., *The Monks of Rabban Hormizd* (repr. Piscataway, NJ: Gorgias Press, 2012).
Vosté, J., 'Les inscriptions de Rabban Hormizd et de Notre-Dame des semences', *Le Muséon* 43 (1930), 263–316.
Walker, J., 'From Nisibis to Xi'an: The Church of the East in Late Antique Eurasia', in *The Oxford Handbook of Late Antiquity* (Oxford: Oxford University Press, 2012), 1–31.
Wallis Budge, E. A., *The Life of Rabban Hormizd and the Foundation of his Monastery at Al-Kosh* (Berlin: Emil Felber, 1894; repr. Piscataway, NJ: Gorgias Press, 2012).
Williams, A., 'Zoroastrians and Christians in Sasanian Iran', *BJRL* 78:3 (1996), 37–53.
Wilmshurst, D., *The Ecclesiastical Organisation of the Church of the East, 1318–1913* (Louvain: Peeters, 2000).
Wilmshurst, D., *The Martyred Church. A History of the Church of the East* (London: East and West Pub., 2011).
Wilmshurst, D., 'The Church of the East in the 'Abassid Era', in *Syriac World*, 189–201.
Wood, P., 'Historiography in the Syriac Speaking World', in *Syriac World*, 405–21.
Yacoub, J., 'La reprise à Chypre en 1445 du nom de "Chaldéens" par les fidèles de l'Église d'Orient', *Istina* 49 (2004), 378–90.
Yacoub, J., *Le Moyen-Orient syriaque: la face inconnue des Chrétiens d'Orient* (Paris: Salvator, 2019).
Zaia, I., *History of the Assyrians from Antiquity to the Fall of Byzantium* (Moscow, 2009) (in Russian).

Byzantine History

Cheynet, J.-C., *Histoire de Byzance* (Paris: Presses Universitaires de France, 2017).
Lemerle, P., *Histoire de Byzance* (Paris: Presses Universitaires de France, 1960).
Meyendorff, J., *Imperial Unity and Christian Divisions: The Church 450–680 A.D* (Crestwood, New York: St Vladimir's Seminary Press, 1989).
Ostrogorsky, G., *History of the Byzantine State*, trans. Joan Hussey, 2nd edn (Oxford: Basil Blackwell, 1968).
Shahid, I., *Byzantium and the Arabs*, 4 vols (Washington, DC: Harvard University Press, 1984–95).

Theology

Bethune-Baker, J. F., *Nestorius and His Teaching. A Fresh Examination of the Evidence* (Cambridge: Cambridge University Press, 1969; repr. Piscataway, NJ: Gorgias Press, 2011).

Brock, S., 'The "Nestorian" Church: A Lamentable Misnomer', *BJRL* 78 (1996), 23–36. <http://noahbickart.fastmail.fm/Academic%20Papers/_Sebastian%20Brock/brock_Nestorian%20Unfortunate%20Minomer.pdf>

Brock, S., 'The Church of the East in the Sasanian Empire Up to the Sixth Century and Its Absence from the Councils in the Roman Empire', in Syriac Dialogue: First Non-official Consultation on Dialogue within the Syrian Tradition, with Focus on the Theology of the Church of the East (Vienna: Pro Oriente, 1996), 69–85.

Brock, S., 'The Christology of the Church in the East in the Synods of the Fifth to Early Seventh Centuries: Preliminary Considerations and Materials', *in E. Ferguson (ed.), Doctrinal Diversity: Varieties of Early Christianity. Recent Studies in Early Christianity (New York: Routledge, 1999), 126, 133.*

Brock, S., 'The Syriac Churches in Ecumenical Dialogue on Christology', in A. O'Mahony (ed.) *Eastern Christianity: Studies on Modern History, Religion and Politics* (London: Melisende, 2004), 44–65.

Brock, S., 'Ecumenical dialogue', in *Gorgias Dictionary*, 136.

Brock, S., 'Les contreverses christologiques en syriaque: contreverses réelles et contreverses imaginées', in F. Ruani (ed.) *Les contreverses religieuses en syriaque* (Paris: Geuthner, 2016), 105–17.

Dickens, M., 'Nestorius Did Not Intend to Argue That Christ Had a Dual Nature, But That View Became Labeled Nestorianism', in S. Danver (ed.), *Popular Controversies in World History: Investigating History's Intriguing Questions* (Santa Barbara, CA: ABC-CLIO, 2010), 145–62.

Famerée, J., 'Éphèse et Nestorius: un malentendu christologique', *Revue Théologique de Louvain*, 39:1 (2008), 3–25. <https://www.persee.fr/doc/thlou_0080-2654_2008_num_39_1_3648>

Gribomont, J., 'Le symbole de la foi de Séleucie-Ctesiphon (410)', in R. H. Fischer (ed.), *A Tribute to Arthur Vööbus: Studies in Early Christian Literature and Its Environment* (Chicago: Lutheran School of Theology at Chicago, 1977), 283–94.

Grillmeier, A., *Christ in Christian Tradition: From the Apostolic Age to Chalcedon (451)* (London: SCM Press, 1975).

Hainthaler T., 'Theological Doctrines and Debates within Syriac Christianity', in *Syriac World*, 377–90.

Halleux de, A., 'Le symbole des évêques perses au synode de Séleucie-Ctésiphon (410)', in G. Wiessner (ed.), *Erkenntnisse und Meinungen*, II = Göttinger Orientforschungen, Reihe Syriaca 17, 1978, 161–90.

Halleux de, A., 'Nestorius: histoire et doctrine', *Irénikon* 66 (1993), 38–51, 163–77.

Lammens, H., 'Les chrétiens à la Mecque à la veille de l'Hégire', *Bulletin de l'Institut français d'archéologie orientale* 14 (Cairo, 1918), 191–230.

McGuckin, J., 'Nestorius and the Political Factions of Fifth-Century Byzantium: Factors in His Personal Downfall', *BJRL* 78:3 (1996), 7–21.
Menze, V., 'The Establishment of the Syriac Churches', in *Syriac World*, 105–18.
Nau, F., Nestorius: Le livre d'Héraclide de Damas (Paris: Letouzey et Ané, 1910; repr. Piscataway, NJ: Gorgias Press, 2010).
Seleznyov, N., 'Nestorius of Constantinople: Condemnation, Suppression, Veneration; with Special Reference to the Role of His Name in East-Syriac Christianity', *Journal of Eastern Christian Studies* 62:3–4 (2010), 165–90.
Seleznyov, N., *The Church of the East and Its Theology. History of Studies*, <https://www.marquette.edu/maqom/hist_stud.pdf>
Tisserand, E., 'Église nestorienne, relations avec Rome', *Dictionnaire de théologie catholique* 11:1 (1931), cols 218–25.
Tisserand, E., 'Église nestorienne, union à Rome, Église chaldéenne catholique', *Dictionnaire de théologie catholique* 11:1 (1931), cols 225–47.
Williams, D., 'The Evolution of Pro-Nicene Theology in the Church of the East', *in From the Oxus River, 387–95.*
Winkler, D., 'The Current Theological Dialogue with the Assyrian Church of the East', in R. Lavenant (ed.), *Symposium Syriacum VII, Orientalia Christiana Analecta 256 (1998), 159–73.*
Winkler, D., 'The Syriac Church Denominations', in *Syriac World*, 119–33.

Schools

Becker, A., *Fear of God and the Beginning of Wisdom. The School of Nisibis and Christian Scholastic Culture in Late Antique Mesoptamia* (Philadelphia, PA: University of Pennsylvania Press, 2006).
Briquel Chatonnet, F., 'La religion comme enseignement. Les écoles dans la tradition historique et culturelle de l'Église syro-orientale', *Comptes-rendus des séances de l'Académie des Inscriptions et Belles-Lettres* (2008), 59–76.
Gutas, D., *Greek Thought, Arabic Culture: The Graeco-Arabic Translation Movement in Baghdad and Early Abbasid Society* (London: Routledge, 1998).
Lacy O'Leary de, G., *How Greek Science Passed to the Arabs* (London: Routledge & Kegan Paul, 1949; repr. London: Routledge, 2016).
Vööbus, A., *History of the School of Nisibis*. CSCO 266/26 (Louvain: Peeters, 1965).

East Syriac Literature

Albert, M., 'Langue et littérature syriaque', in M. Albert et al. (eds), *Christianismes orientaux* (Paris: Éd. du Cerf, 1993).
Beulay, R., *La lumière sans forme. Introduction à l'étude de la mystique chrétienne syro-orientale* (Chevetogne: Éd de Chevetogne, 1987).

Pinggéra, K. et G. Kessel, *A Bibliography of Syriac Ascetic and Mystical Literature* (Louvain: Peeters, 2011).
Pirtea, A., 'The Mysticism of the Church of the East', in *Syriac World*, 355–76.
Teule, H. and V. Schepens, 'Christian Arabic Bibliography 1995–2000', Journal of Eastern Christian Studies 58 (2006), 265–300.
Treiger, A., Christian Arabic: A Classified Bibliography for Researchers <https://www.academia.edu/1971015/Christian_Arabic_A_Unified_Bibliography_2000-2012_>; voir aussi <http://syri.ac/christianarabic>.

Chapter 2: In Arabia and the Persian Gulf

Bernard, V. et al., 'L'église d'al-Qousour Failaka, État de Koweit', *Arabian Archaeology and Epigraphy* 2 (1991), 145–81.
Bowman, J., 'The Sasanian Church in the Kharg Island', *Acta Iranica* 1 (Tehran, 1974), 217–20.
Bowman, J., 'Christian monastery on the Island of Kharg', *Australian Journal of Biblical Archaeology* 2:3 (1975), 49–64.
Brelaud, S., 'Al-Hira et ses chrétiens dans les guerres romano-perses', Camenulae 15 (2016), 1–26.
Calvet, Y., 'Monuments paléo-chrétiens à Koweit et dans la région du Golfe, Symposium Syriacum, Uppsala University, Department of Asian and African Languages, 11–14 August 1996', *Orientalia Christiana Analecta* 256 (1998), 671–85.
Calvet, Y., 'Vestiges de l'expansion nestorienne dans le Golfe', *Le Monde de la Bible* 119 (1999), 25–7.
Carter, R., 'Christianity in the Gulf during the First Centuries of Islam', *Arabian Archaeology and Epigraphy* 19 (2008), 71–108.
Chiesa, B. et al. (eds), *L'Arabie avant l'Islam* (Aix-en-Provence: Edisud, 1994).
Elders, J., 'The Lost Churches of the Arabian Gulf: Recent Discoveries on the Islands of Sir Bani Yas and Marawah, Abu Dhabi Emirate, United Arab Emirates', *Proceedings of the Seminar for Arabian Studies* 31 (2001), 47–58.
Elders, J., *The Nestorians in the Gulf: Just Passing Through? Recent Discoveries on the Island of Sir Bani Yas, Abu Dhabi Emirate, UAE* (London: Trident Press, 2003), 230–6.
Fiey, J. M., *Assyrie chrétienne*. Vol. 3, *Bét Garmaï, Bét Aramayé et Maishan nestoriens* (Beirut: Dar El-Machreq Éditeurs, 1968).
Fiey, J. M., 'Les diocèses syriens orientaux du Golfe Persique', in F. Graffin (ed.), *Mémorial Mgr. Gabriel Kouri-Sarkis* (Louvain : Imprimerie Orientaliste, 1969), 177–219.
Fujii, H. et al., 'Excavations at Ain Sha'ia Ruins and Dukakin Caves', *Al Rafidan* 10 (1989), 27–88.
Gachet, J. 'Akkaz (Kuwait), a Site of the Partho-Sasanian period', Proceedings of the Seminar for Arabian Studies 28 (1998), 69–79.
Healey, J. F., 'The Christians of Qatar in the 7th Century A.D.', in I. R. Netton (ed.), *Studies in Honour of Clifford Edmund Bosworth*. Vol. 1: *Hunter of the East: Arabic and Semitic Studies* (Leiden: Brill, 1999), 222–37.

Healey, J. F., 'The Patriarch Išo'yabh and the Christians of Qatar in the First Islamic Century', in *Christian Heritage*, 1–9.

Hellyer, P., 'Nestorian Christianity in Pre-Islamic UAE and Southeastern Arabia', Journal of Social Affairs 18:72 (2001), 79–92, and in Bibliotheca Hagiographica Orientalis, 527–30. <http://www.adias-uae.com/publications/hellyer01b.pdf>

Hunter, E., 'Report and Catalogue of Inscribed Fragments: Ain Sh'a and Dukakin Caves Near Najaf, Iraq', repr. from *Al-Rafidan*, vol. 10 (Tokyo: The Institute for Cultural Studies of Ancient Iraq, Kokushikan University, 1989), 89–107.

Hunter, E., 'An Inscribed Reliquary from the Middle Euphrates', *Oriens Christianus* 75 (1991), 147–65.

Hunter, E., 'Syriac Inscriptions of al-Hira', *Oriens Christianus* 80 (1996), 66–81.

Hunter, E., 'Christian Matrix of al-Hira', in C. Jullien (ed.), *Chrétiens en terre d'Iran*. Vol. 2 *Contreverses des chrétiens dans l'Iran sassanide* (Paris: Association pour l'avancement des études iraniennes, 2008), 41–56.

Jullien, C. and F. Jullien, 'Le monachisme dans le Golfe persique, six siècles d'histoire', in M. J. Steve, et al. (ed.), *L'île de Kharg. Une page de l'histoire du Golfe Persique et du monachisme oriental* (Neuchâtel: Recherches et Publications, 2003), 155–83.

Jullien, F., 'La réforme d'Abraham de Kashkar dans le golfe Persique: Le monastère de l'Île de Kharg', *Parole de l'Orient* 31 (2006), 201–11.

King, G., 'A Nestorian Monastic Settlement on the Island of Sir Bani Yas, Abu Dhabi', *Bulletin of the School of Oriental and African Studies* 60 (1997), 221–35.

King, G., D. Dunlop et al., 'A Report on the Abu Dhabi Islands Archaeological Survey (1993–1994)', in *Proceedings of the Seminar for Arabian Studies* 25 (1995), 63–74.

Kozah, M. et al. (eds), *The Syriac Writers of Qatar in the Seventh Century*. Gorgias Eastern Christian Studies 38 (Piscataway, NJ: Gorgias Press, 2014).

Langfeldt, J. A., 'Recently Discovered Early Christian Monuments in Northeastern Arabia', Arabian Archaeology and Epigraphy 5 (1994), 32–60.

Matsumoto, K., 'Dukakin caves', in H. Fuji et al. (eds), 'Excavations at Ain Sha'ia Ruins and Dukakin Caves', *Al-Rafidan* 10 (1989), 81–5.

Missick, S., 'Socotra: The Mysterious Island of the Church of the East', *JAAS* 16:1 (2002), 96–109.

Okada, Y., 'Early Christian Architecture in the Iraqi South-Western Desert', *Al-Rafidan* 12 (1991), 71–83.

Okada, Y. and H. Numoto, 'Fortified Building-Site F', in H. Fuji et al. (eds), 'Excavations at Ain Sha'ia Ruins and Dukakin Caves', *Al-Rafidan* 10 (1989), 35–61.

Oussani, G., 'Arabia', *The Catholic Encyclopedia*, vol. 1 (New York, 1907) <http://www.newadvent.org/cathen/01663a.htm>

Payne, R., 'Monks, Dinars and Date Palms: Hagiographical Production and the Expansion of Monastic Institutions in the Early Islamic Persian Gulf', *Arabian Archaeology and Epigraphy* 22 (2011), 97–111.

Pereira, F., 'La Chrétienté de l'île de Socotora', *Aethiops* 2:1 (January 1923), 1–4.

Potts, D. T. (ed.), *Dilmun. New Studies in the Archaeology and Early History of Bahrain*, (Berlin: D. Reimer Verlag, 1983).

Potts, D. T., *Nestorian Crosses from Jabal Berri* (Sydney, 1994) <https://onlinelibrary.wiley.com/doi/abs/10.1111/j.1600-0471.1994.tb00055.x>

Rice, D. T., 'Hira', *Journal of the Royal Asiatic Society* 19 (1932), 25–68.
Rice, D. T., 'The Oxford Excavations at Hira. 1931', *Antiquity* 6 (1932), 76–91.
Robin, C., 'Arabia and Ethiopia', in S. Fitzgerald Johnson (ed.), *The Oxford Handbook of Late Antiquity* (Oxford: Oxford University Press, 2012), 247–332.
Rompay van, L., 'Najran', in *Gorgias Dictionary*, 302–3.
Salles, J. F., 'Chronologies du monachisme dans le golfe arabo-persique', in F. Jullien and M.-J. Pierre (eds), *Les Monachismes d'Orient* (Turnhout: Brepols, 2011), 291–312.
Salles, J. F. and O. Callot, 'Les églises antiques de Koweit et du golfe Persique ', in Briquel 2013, 237–68.
Simpson, J., 'Christians in Iraq's Desert Frontier', *Al-Rafidan* 19 (2018), 1–30.
Steve, M. J. et al., *L'île de Kharg. Une page de l'histoire du Golfe persique et du monachisme oriental*. Civilisation du Proche-Orient 1 (Neuchâtel: Recherches et Publications, 2003).
Takahashi, H., 'Hirta', *Gorgias Dictionary*, 198–9.
Trimingham, J. S., *Christianity among the Arabs in Pre-Islamic Times* (London: Longmans, Green & Co., 1979; repr. London: Stacey Publishing, 1990).
Vincent B. and J.-F. Salles, 'Discovery of a Christian Church at Al-Qusur, Failaka (Kuwait)', Proceedings of the Seminar for Arabian Studies 21 (1991), 7–21.
Vincent B. et al., 'L'église d'al-Qousour Failaka, État de Koweit', Arabian Archaeology and Epigraphy 2 (1991), 145–81.

Chapter 3: In India

Anon., 'The Saint Abimalek Timotheus Metropolitan (1908–1945)', *Voice of the East* 66:7–8, (July–August 2019).
Aprem (Mar), 'Church of the East', in G. Menchery and E. Hambye (eds) *St. Thomas Christian Encyclopaedia of India* (Thrissur: St Thomas Christian Encyclopedia of India, 1973).
Aprem (Mar), *Mar Thoma Darmo (A Biography)* (Thrissur: Mar Sliwa Church, 1974).
Aprem (Mar), *Mar Abimalek Timotheus*: a Biography (Thrissur: Mar Timotheus Memorial Orphanage, 1975).
Aprem (Mar), *The Chaldean Syrian Church in India* (Thrissur: Mar Narsai Press, 1977).
Aprem (Mar), *Mar Abdisho Thondanat* (Thrissur: Mar Narsai Press, 1987).
Aprem (Mar), 'Mar Narsai Press', *Bulletin of the John Rylands Library* 78:3 (1996), 171–8.
Aprem (Mar), The Assyrian Church of the East in the Twentieth Century (Kottayam: SEERI, 2003).
Briquel Chatonnet, F., 'Syriac Manuscripts in India, Syriac Manuscripts from India', Hugoye: Journal of Syriac Studies 15:1 (2012), 281–91.
Brock, S. (ed.), 'The Syriac Churches in India', in *Hidden Pearl*, vol. 3, 13–14.
Brock, S., 'Chaldean Syrian Church', in *Gorgias Dictionary*, 92–3.
Brock, S., 'Thomas Christians', in *Gorgias Dictionary*, 410–14.
Brown, L., *The Indian Christians of Saint Thomas* (Cambridge: Cambridge University Press, 1956; repr. 1982).

Buchanan, C., *Christians Researches in Asia* (London: Henry Washbourne, 1819); see the chapter 'Syrian Christians in India', 104–42.
Cannuyer, C., 'Vingt siècles de Christianisme en Inde méridionale', in *Solidarité Orient* (Bruxelles, 2004), 4–20.
Cosmas Indicopleustès, *Topographie chrétienne*, ed. and trans. W. Wolska-Conus, 3 vols (Paris: Sources chrétiennes, 1968–73).
Frykenberg, R., *Christianity in India: From Beginnings to the Present* (Oxford: Oxford University Press, 2008).
Hambye, E., *Eastern Christianity in India:* A History of the Syro-Malabar Church from the Earliest Time to the Present Day (London: Longmans, Green & Co., 1957).
Ignatius, K. A., *The Holy Apostolic Catholic Assyrian Church of the East. A Short History* (Thrissur, 2010).
Ignatius Yacoub III, *History of the Syrian Church of India*; trans. M. Moosa (repr. Piscataway, NJ: Gorgias Press, 2009).
Kollaparambil, J., *The Archdeacon of All-India* (Rome: Pontificia Universitas Lateranensis, 1972).
Mackenzie, G. T., *Christianity in Travancore* (Piscataway, NJ: Gorgias Press, 2011).
Malekandathil, P., *Maritime India: Trade, Religion and Polity in the Indian Ocean* (New Delhi: Primus Books, 2010).
Medlycott, A., *India and the Apostle Thomas* (Piscataway, NJ: Gorgias Press, 2005).
Menachery, G. and E. Hambye (eds), *St. Thomas Christian Encyclopaedia of India*, Trichur, 2 vols (Trichur: St Thomas Christian Encyclopedia of India, 1973–82).
Mingana, A., *The Early Spread of Christianity in India* (Manchester: repr. from *BJRL*, 1925; repr. Piscataway, NJ: Gorgias Press, 2010).
Mundadan, A. M., *History of Christianity in India*, vol. 1, *From the Beginning up to the Middle of the Sixteenth Century* (Bangalore: Church History Association of India, 1984).
Nedungatt, G., *Quest* for the Historical. *Thomas, Apostle* of India (Bangalore: Theological Publications in India, 2008).
Neill, S. (ed.), *A History of Christianity in India. The Beginning to AD 1707*, 2 vols (Cambridge: Cambridge University Press, 1984–5).
Perczel. I., 'Syriac Christianity in India', in *Syriac World*, 653–97.
Puthur, B., *The Life and Nature of the St Thomas Christian Church in the Pre-Diamper Period* (Kochi: Liturgical Research Centre of the Syro-Malabar Church, 2000).
Scher, A. (ed. and trans.), 'Histoire nestorienne inédite: Chronique de Séert', Part I, *Patrologia Orientalis* 4:3 (1908), 5:2 (1910) ; Part II, *Patrologia Orientalis* 7:2 (1911), 13:4 (1919).
Takahashi, H., 'Diamper, Synod of', *Gorgias Dictionary*, 118–19.
Thekeparampil, J., 'Vestiges of East Syriac Christianity in India', in *Jingjiao*, 485–98.
Thekeparampil, J. et al., 'Les églises des chrétiens de saint Thomas au Kérala', in Briquel 2013, 467–90.
Tisserand, E., 'Église syro-malabare', *Dictionnaire de théologie catholique* 14:2 (1941), cols 3089–162.
Tubach, J., 'The Mission Field of the Apostle Thomas', in *Hidden Treasures*, 291–303.
Valavanthara, A., *India in 1500 AD: The Narratives of Joseph the Indian* (Piscataway, NJ: Gorgias Press, 2010).

Varghese, B., 'East Syrian Missions to the Malabar Coast in the Sixteenth Century', in *From the Oxus River*, 317–40.

Wood, P., *The Chronicle of Seert: Christian Historical Imagination in Late Antique Iraq* (Oxford: Oxford University Press, 2013).

Chapter 4: In Central Asia and Beyond. On the Silk Road

Aprem (Mar), *Nestorian Missions* (Trichur: Mar Narsai Press, 1976; repr. 2002).

Ashurov, B., 'Tarsakya: an Analysis of Sogdian Christianity based on archaeological, numismatic, epigraphic and textual sources' (thesis, SOAS, University of London, 2013) <http://eprints.soas.ac.uk/18057/1/Ashurov_3569.pdf>

Ashurov, B., 'Inculturation matérielle de l'Église d'Orient en Asie centrale: témoignages archéologiques', Études syriaques 12, 161–84.

Ashurov, B., 'Sogdian Christian Texts: Socio-Cultural Observations' <https://www.academia.edu/12196132/Sogdian_Christian_Texts_Socio-Cultural_Observations>

Asimov, M. and C. E. Bosworth (eds), *History of Civilizations of Central Asia*, 6 vols (Paris: UNESCO, 1998), 1992–2005.

Asmussen, J. P., 'The Sogdian and Uighur-Turkish Christian Literature in Central Asia Before the Real Rise of Islam: A Survey', in L. A. Hercus et al. (eds), *Indological and Buddhist Studies: Volume in Honour of Professor J. W. de Jong on His Sixtieth Birthday* (Canberra: Australian National Faculty of Asian Studies, 1982), 11–29.

Atwood, P. 'Historiography and Transformation of Ethnic Identity in the Mongol Empire: The Öngüt Case', *Asian Ethnicity* 15:4 (2014), 516–34.

Bardaisan, *Book of the Laws of Countries*, ed. F. Nau, *Patrologia Syriaca* 1:2 (Paris, 1907), 607.

Bardesane, *Le Livre des Lois* des Pays; trans. P.-H. Poirier (Paris: Les Belles Lettres, 2020).

Baumer, C., *Southern Silk Road* (Bangkok: Orchid Press, 2000).

Baumer, C., 'Survey of Nestorianism and of Ancient Nestorian Architectural Relics in the Iranian Realm', in *Jingjiao*, 445–74.

Baumer, C., *Traces in the Desert: Journeys of Discovery across Central Asia* (London and New York: I. B. Tauris, 2008).

Baumer, C., *The History of Central Asia*, vol. 2, *The Age of the Silk Roads*; vol. 4, *The Age of Decline and Revival* (London: I. B. Tauris, 2014 and 2018).

Baumer, C., 'The Mission to the East', and 'Recent Archaeological Discoveries and Ecclesiastical and Political Developments', in Baumer 2016, 169–94, 287–93.

Beckwith, C., *Empires of the Silk Road: A History of Central Eurasia from the Bronze Age to the Present* (Princeton: Princeton University Press, 2009).

Blochet, E., 'La conquête des états nestoriens de l'Asie Centrale par les Shiites', *Revue de l'Orient chrétien* 25 (1925–6), 3–132.

Bonavia, J. and C. Baumer, *The Silk Road: Xian to Kashgar* (Hong Kong: Odyssey Books, 2004).

Borbone, P. G., 'Syroturcica. 1, The Önggüds and the Syriac Language', in G. A. Kiraz (ed.), *Malphono w-Rabo d-Malphone: Studies in Honor of S. P. Brock* (Piscataway, NJ: Gorgias Press, 2008), 1–17.

Borbone, P. G., 'Les églises d'Asie centrale et de Chine: état de la question à partir des textes et des découvertes archéologiques: essai de synthèse', in Briquel 2013, 441–65.

Borbone, P. G., 'An Önggüd Gravestone in the Musée Guimet, Paris, and Its Inscription', in P. Fedi and M. Paolillo (eds), *Arte dal Mediterraneo al Mar della Cina. Genesi ed incontri di scuole e stili. Scritti in onore di Paola Mortari Vergara Caffarelli* (Palermo: Officina di studi medievali, 2015), 221–30.

Borbone, P. G. and P. Marsone (eds), *Le christianisme syriaque en Asie centrale et en Chine* (Paris: Geuthner, 2015).

Boulnois, L., *La route de la Soie: dieux, guerriers et marchands* (Geneva: Olizane Éds, 2010).

Brock, S., 'Bar Shabba/Mar Shabbay, First Bishop of Merv', in M. Tamcke et al., *Syrisches Christentum weltweit. Studien zur syrischen Kirchengeschichte. Festschrift für Prof. Hage* (Münster, 1995), 190–201.

Christian, D. and C. Benjamin (eds), *Realms of the Silk Roads: Ancient and Modern*. Silk Roads Studies 4 (Turnhout: Brepols, 2000).

Christian, D. and C. Benjamin (eds), *Walls and Frontiers in Inner-Asian History*. Silk Roads Studies 6 (Turnhout: Brepols, 2002).

Chwolson, D., Syrisch-Nestorianische Grabinschriften Aus Semirjetschie (St Petersburg, 1897; repr. Piscataway, NJ: Gorgias Press 2013).

Colless, B., 'The Nestorian Province of Samarqand', *Abr Nahrayn* 24 (1986), 51–7.

Comneno, L., 'Nestorianism in Central Asia during the First Millennium: Archaeological Evidence', *Journal of the Assyrian Academic Society* 11:1 (1997), 20–69.

Dauvillier, J., 'Les provinces chaldéennes de l'extérieur au Moyen-Âge', *Mélanges offerts au R. P. Ferdinand Cavallera* (Toulouse: Bibliothèque de l'Institut catholique, 1948), 260–316.

Dauvillier, J., 'Byzantins d'Asie Centrale et d'Extrême-Orient au Moyen-Âge', *Revue des Études Byzantines* 11 (1953), 62–87.

Dauvillier, J., 'L'expansion de l'Église syrienne en Asie Centrale et en Extrême Orient', *L'Orient syrien* 1 (1956), 76–87.

Desreumaux, A., 'La collection des pierres tombales syro-orientales du Turkestan conservées à Paris et à Lyon', in *Études syriaques* 12, 237–56.

Dickens, M., 'Syriac Gravestones in the Tachkent History Museum', in *Hidden Treasures*, 13–49.

Dickens, M., 'The Syriac Bible in Central Asia', in *Christian Heritage*, 92–120.

Dickens, M., 'Patriarch Timothy I and the Metropolitan of the Turks', *Journal of the Royal Asiatic Society* 20:2, 117–39. <https://www.academia.edu/436102/Patriarch_Timothy_I_and_the_Metropolitan_of_the_Turks>

Dickens, M., 'A Christian Monastic Community in Medieval Central Asia: Evidence from Syriac Inscriptions Near Urgut, Uzbekistan', *Middle East Studies Association Annual Conference* (San Diego, CA), 18–21 November 2010.

Dickens, M., 'Le christianisme syriaque en Asie Centrale', in *Études syriaques* 12, 5–39.

Dickens, M., 'More Gravestones in Syriac Script from Tashkent, Panjikent and Asganat', in *Winds of Jingjiao*, 105–29.

Dickens, M., 'Syriac Inscriptions Near Urgut, Uzbekistan', *Studia Iranica* 46 (2017), 205–60 <http://www.exploration-eurasia.com/pictures/Urgut_inscriptions.pdf>

Dickens, M., 'Syriac Christianity in Central Asia', in *Syriac World*, 583–624.
Dickens, M., 'Nestorian Christianity in Central Asia' <https://www.academia.edu/398258/Nestorian_Christianity_In_Central_Asia>
Dickens, M. and A. Savchenko, 'Prester John's Realm: New Light on Christianity between Merv and Turfan', in *Christian Heritage*, 121–35.
Di Cosmo, N., The Cambridge History of Inner Asia (Cambridge: Cambridge University Press, 2015).
Fisher, W. B. et al. (eds), *The Cambridge History of Iran* (Cambridge: Cambridge University Press), 1993–2003.
Foltz, R., *Religions of the Silk Road: Overland Trade and Cultural Exchange from Antiquity to the Fifteenth Century* (Basingstoke: Macmillan, 2000).
Frankopan, P., *The Silk Roads: A New History of the World* (London: Bloomsbury Publishing, 2015).
Frankopan, P., *The New Silk Roads: The Present and Future of the World* (London: Bloomsbury Publishing, 2018).
Frye, R., *The Heritage of Central Asia from Antiquity to the Turkish Expansion* (Princeton: Princeton University Press, 1996).
Grousset, R., *L'Empire des Steppes. Attila, Gengis-Khan, Tamerlan* (Paris, 1939; repr. Paris: Payot, 2001).
Halévy, J., 'De l'introduction du christianisme chez les tribus turques de la Haute Asie', Revue de l'Histoire des Religions 22 (1890), 289–301.
Hansen, O, 'Die christliche Literatur der Sogdier', in B. Spuler (ed.) *Handbuch der Orientalistik*, I/IV: 2.1 (Leiden and Cologne: Brill, 1968), 91–9.
Hansen, V., 'The Impact of the Silk Road Trade on a Local Community: The Turfan Oasis, 500–800', in E. Vaissière and E. Trombert (eds), *Les Sogdiens en Chine* (Paris: École française d'Extrême-Orient, 2005), 283–310.
Hansen, V., *The Silk Road: A New History* (Oxford: Oxford University Press, 2012).
Hunter, E., 'The Conversion of the Kerait to Christianity in A.D. 1007', Zentralasiatische Studien 22 (1989–91), 142–63.
Hunter, E., 'Syriac Christianity in Central Asia', Zeitschrift für Religions und Geistesgeschichte 44 (1992), 362–8.
Hunter, E., 'The Church of the East in Central Asia', BJRL 78 (1996), 129–42.
Hunter, E., 'Converting the Turkic Tribes', in C. Benjamin and S. Lieu (eds), *Walls and Frontiers in Inner-Asian History*. Silk Road Studies 6 (Turnhout: Brepols, 2002), 183–95.
Huyghe, E. and F.-B. Huyghe, *La route de la Soie ou les empires du mirage* (Paris: Payot & Rivages, 2017).
Johnson, S., 'The Westwardness of Things: Literary Geography and the Church of the East', in *Winds of Jingjiao*, 183–202.
Juliano, A. and J. Lerner (eds), *Monks and Merchants: Silk Road Treasures from Northern China. Gansu and Ningxia Fourth-Seventh Century* (New York: Harry N. Abrams, 2001).
Khorikyan, H., 'On the Location of the Hephtalites', Journal of Oriental Studies 14 (2018), 472–86.
Klein, W., 'A Christian Heritage on the Northern Silk Road: Archaeological and Epigraphic Evidence of Christianity in Kyrgystan', Journal of the Canadian Society for Syriac Studies 1 (2001), 85–100.

Klein, W., 'Les inscriptions syriaques d'Asie centrale', in Briquel 2004, 125–41.
Klein, W., 'A Newly Excavated Church of Syriac Christianity along the Silk Road in Kyrghyzstan', *The Journal of Eastern Christian Studies* 56 (2004), 25–35.
Klein, W., 'Zwei Neu Gefundene Syrische Grabsteine aus Kyrgyzstan', in Hidden Treasures, 87–90.
Klimkeit, H. J., *Christian Art on the Silk Road* (1993).
Koshelenko, G. et al., 'The Beginnings of Christianity in Merv', *Iranica Antiqua* 30 (1995), 60–70.
Kyzlasov, L., 'Archaeological Research in Aq-Beshim', *Trudy Kygizskoi arkheologo-etnograficheskoi ekspeditcii* 2 (1959), 155–227.
Lerner, J., 'The Merchant Empire of the Sogdians', in A. Juliano and J. Lerner (eds), *Monks and Merchants: Silk Road Treasures from Northern China. Gansu and Ningxia Fourth–Seventh Century* (New York: Harry N. Abrams, 2001), 220–30.
Litvinsky, B., 'The Hephtalites Empire', in B. Litvinsky (ed.), *History of the Civilizations of Central Asia* 3 (Paris: UNESCO, 1996), 135–62.
Liu, X., *The Silk Road in World History* (Oxford: Oxford University Press, 2010).
Mackerras, C., 'The Uighurs', in D. Sinor (ed.), *The Cambridge History of Early Inner Asia* (Cambridge: Cambridge University Press, 1990), 317–42.
Mingana, A., *The Early Spread of Christianity in Central Asia and the Far East* (Manchester, repr. from *BJRL*, 1925; repr. Piscataway, NJ: Gorgias Press, 2010).
Nau, F., 'L'expansion nestorienne en Asie', *Annales du Musée Guimet* 40 (1914), 193–383.
Nau, P., 'Les pierres tombales nestoriennes du Musée Guimet', Revue de l'Orient chrétien 18 (1913), 3–35.
Naymark, A., 'Christians in Pre-Islamic Bukhara: Numismatic Evidence', in J. Elverskog and A. *Naymark* (eds), *Annual Central Eurasian Studies Conference, 1994–1996* (Bloomington, 1996).
Naymark, A., 'Christian Principality in the Bukharan Oasis', *Journal of Oriental Numismatic Society* 206 (2011), 2–3.
Nicolini N. and R. Malek, 'Preliminary Bibliography on the Church of the East in China and in Central Asia', in *Jingjiao*, 499–698.
Nicolini-Zani, M., 'Christiano-Sogdica: An Updated Bibliography on the Relations between Sogdians and Christians throughout Central Asia and into China', in M. Comparetti et al. (eds), *Studies Presented to Boris Ilich Marshak on the Occasion of His 70th Birthday* (Venice, 2006), 455–70 <https://www.transoxiana.org/Eran/Articles/nicolini-zani.html>
Paolillo, M., 'White Tatars: The Problem of the Origin of the Öngüt Conversion to *Jingjiao* and the Uighur Connection', in *From the Oxus River*, 237–54.
Parry, K., 'Byzantine-Rite Christians (Melkites) in Central Asia and China and their Contacts with the Church of the East', in *Winds of Jingjiao*, 203–20.
Pelliot, P., 'Chrétiens d'Asie Centrale et d'Extrême Orient', *T'oung Pao* 15 (1914), 623–44.
Platt, A., 'Changing Mission at Home and Abroad: Catholicos Timothy I and the Church of the East in the Early Abbasid Period', in *Winds of Jingjiao*, 161–82.
Rott, P., 'Christian Crosses from Central Asia', in *Jingjiao*, 395–401.
Roux, J.-P., *L'Asie Centrale, Histoire et civilisations* (Paris: Fayard, 1997) (see the maps).

Ruji, N., 'A Nestorian Tombstone with Syriac Inscriptions from Central Asia', in *From the Oxus River*, 93–8.
Ryan, J., 'Preaching Christianity along the Silk Road: Missionary Outposts in the Tartar Middle Kingdom in the Fourteenth Century', *Journal of Early Modern History* 2 (1998), 350–73.
Savchenko, A., 'Urgut Revisited', ARAM, *Journal of the Society for Syro-Mesopotamian Studies* 8:1–2 (Oxford, 1996), 333–54 <http://www.iranicaonline.org/articles/urgut>
Savchenko, A., *The Monastery of Urgut Investigated by the East Sogdian Archaeological Expedition* (Kiev: ESAREX, 1999) <htpp://www.apple.kiev.ua.esarex>
Savchenko, A., 'Östliche Urkirche in Usbekistan', in P. von Zabern (ed.), *Antike Welt* 2 (2010), 74–82.
Semenov, G., Studien zur sogdischen Kultur an der Seidenstrasse, *Studies in Oriental Religions* 36 (1996).
Semenov, G., 'History of Archaeological Research of Aq-Beshim', in *Sūyāb. Aq-Beshim*, Archaeological Expedition of the State Hermitage Museum (St Petersburg, 2002), 4–10.
Sims-Williams, N., 'Sogdian Translations of the Bible', in E. Yarshater (sd.), *Encyclopaedia Iranica* 4:2 (London: Routledge, 1990).
Sims-Williams, N., 'Christianity iii. In Central Asia and Chinese Turkestan', *Encyclopædia Iranica* 5:5 (1991), 530–4 <http://www.iranicaonline.org/articles/christianity-iii>
Sims-Williams, N., 'The Sogdian Inscriptions of Ladakh', in K. Jettmar, et al. (eds), *Antiquities of Northern Pakistan. Reports and Studies*, vol. 2 (Mainz, 1993), 151–63.
Sims-Williams, N., 'Christianity in Iran and Central Asia', in R. Asher (ed.), *The Encyclopedia of Language and Linguistics* (Oxford, 1994), 548–9.
Sims-Williams, N., 'The Sogdian Merchants in China and India', in A. Cadonna and L. Lanciotti (eds), *Cina ed Iran da Alessandro Magno alla dinastia Tang* (Florence, 1996), 45–67.
Sims-Williams, N., 'Christian Literature in Middle Iranian Languages', in R. Emmerick et al. (eds), *The Literature of Pre-Islamic Iran*. Companion vol. 1 to *A History of Persian Literature* (London: I. B. Tauris, 2009), 266–87.
Sinor, D. (ed.), *The Cambridge History of Early Inner Asia* (Cambridge: Cambridge University Press, 1990).
Stewart J., *Nestorian Missionary Enterprise: The Story of the Church on Fire, 1928* (New York, repr. Thrissur: Mar Narsai Press, 1961).
Tang, L., 'A History of Uighur Religious Conversions (5th–16th Centuries)' <https://www.academia.edu/3022045/A_History_of_Uighur_Religious_Conversions_5th_-_16th_Centuries_2005_>
Tang, L., 'Turkic Christians in Central Asia and China (5th–14th Centuries)' <https://www.academia.edu/7158958/_Turkic_Christians_in_Central_Asia_and_China_5th_-_14th_Centuries_In_Memory_of_Prof._Geng_Shimin_?email_work_card=title>
Timothée Ier, *Lettres*, in O. Braun (trans.), CSCO 74/75 (1914–15).
Tisserand, E., 'Église nestorienne en Asie centrale', *Dictionnaire de théologie catholique* 11:1 (1931), cols 207–13.

Toepel, A., 'Christians in Korea at the End of the Thirteenth Century', in *Hidden Treasures*, 279–89.
Tredinnick, J., *Xinjiang. China's Central Asia* (Hong Kong: Odyssey Books, 2012).
Tucker, J., *The Silk Road:* Art and History (London: Philip Wilson, 2003) (see the maps).
Tucker, J., *The Silk Road: Central Asia, Afghanistan and Iran: A Travel Companion* (London: I. B. Tauris, 2015).
Vaissière, E., *Sogdian Traders: A History*. Handbook of Oriental Studies, Part 8: Uralic Studies and Central Asia (Brill, 2005).
Vaissière, E., 'Sogdiana iii. History and Archeology', *Encyclopædia Iranica*, online edition, 2011, <http://www.iranicaonline.org/articles/sogdiana-iii-history-and-archeology>
Vaissière, E., 'Central Asia and the Silk Road', *The Oxford Handbook of Late Antiquity* (Oxford: Oxford University Press, 2012), 142–69.
Winkler, D. and L. Tang (eds), *Hidden Treasures and Intellectual Encounters: Studies on East Syriac Christianity in China and Central Asia* (Münster: LIT Verlag, 2009).
Yakubovich, 'Sogdian', in *Gorgias Dictionary*, 382–3.
Yarshater (ed.), *The Cambridge History of Iran*, vol. 3, chapter 6, 1983, 232–62.
Zeymal, E., 'The Political History of Transoxiana', in E. Yarshater (ed.), *The Cambridge History of Iran*, vol. 3, chapter 6, 1983, 232–62.
Zieme, P., 'Notes sur les textes chrétiens en vieux-ouïghour', in Études syriaques 12, 185–98.

Chapter 5: In China under the Tang (635–845) and the Mongols (1206–1368)

Aprem (Mar), 'Reference to China in Syriac Sources', in *Hidden Treasures*, 183–93.
Barbati, C., 'La documentation sogdienne chrétienne et le monastère de Bulayïq', in *Études syriaques* 12, 89–120.
Barrett, T., 'Buddhism, Taoism and Chinese Christianity', *Bulletin of the School of Oriental and African Studies* 65:3 (2002), 555–60.
Borbone, P. G., 'I blochi con croci e iscrizione siriaca da Fangshan', *Christiana Orientalia Periodica* 72 (2006), 167–87.
Borbone, P. G., 'Syroturcica. 1, The Onggüds and the Syriac Language', in G. A. Kiraz (ed.), *Malphono w-Rabo d-Malphone: Studies in Honor of Sebastian P. Brock* (Piscataway, NJ: Gorgias Press, 2008), 1–17.
Borbone, P. G., 'A 13th-Century Journey from China to Europe: The "Story of Mar Yahballaha and Rabban Sauma"', *Egitto e Vicino Oriente* 31 (2008), 221–42.
Borbone, P. G., 'More on the Priest Särgis in the White Pagoda: The Syro-Turkic Inscription on the White Pagoda in Hohhot', in From the Oxus River, 51–65.
Brock, S., 'Syriac Christianity in China', *The Hidden Pearl*, 2, 196–8.
Brock, S., 'Turfan, Syriac texts from', in *Gorgias Dictionary*, 420–1.

Bundy, D. D., 'Missiological Reflections on Nestorian Christianity in China during the Tang Dynasty', in F. K. Flinn and T. Hendricks (eds), *Religion in the Pacific Era* (New York, NY: Paragon House 1985), 14–30.
Childers, J. W., 'Xian', in *Gorgias Dictionary*, 428–9.
Chonglin, L. and N. Ruji, 'The Discovery of Nestorian Inscriptions from Almaliq, Xinjiang, China', in *Hidden Treasures*, 91–9.
Deeg, M., 'Towards a New Translation of the Chinese Nestorian Documents from the Tang Dynasty', in *Jingjiao*, 115–32.
Deeg, M., 'La littérature chrétienne orientale sous les Tang: un bref aperçu', in *Études syriaques* 12, 199–214.
Dickens, M., 'Multilingual Christian Manuscripts from Turfan', *Journal of the Canadian Society for Syriac Studies* 9 (2009), 22–42.
Dickens, M., 'Scribal Practices in the Turfan Christian Community', *Journal of the Canadian Society for Syriac Studies* 13 (2013), 32–52.
Dickens, M., 'The Importance of the Psalter at Turfan', in *From the Oxus River*, 357–80.
Dickens, M. and N. Sims-Williams, 'Christian Calendrical Fragments from Turfan', in J. Ben-Dov, W. Horowitz and J. M. Steele (eds), *Living the Lunar Calendar* (Oxford: Oxbow Books, 2012), 269–96.
Drake, F. S., 'Nestorian Crosses and Nestorians in China under the Mongols', *Journal of the Hong Kong Branch of the Royal Asiatic Society* 2 (1962), 11–25.
Enoki K., 'The Nestorian Christianity in China in Medieval Time According to Recent Historical and Archaeological Researches', in *L'Oriente Cristiano nella storia della civiltà* (Rome: Accademia Nazionale dei Lincei, 1964), 44–77.
Fairbank, J., D. Twitchett et al. (eds), *The Cambridge History of China*, 15 vols (Cambridge: Cambridge University Press, 1978–2016).
Farina, M., 'A Database of the Syriac and Syro-Turkic Inscriptions from Central Asia and China', in *From the Oxus River*, 67–81.
Ferreira, J., *Early Chinese Christianity: The Tang Christian Monument and Other Documents* (Strathfield, Australia: St Pauls Publications, 2014).
For the text of the Xi'an Stele, see: <https://sourcebooks.fordham.edu/halsall/eastasia/781nestorian.asp;https://archive.org/stream/nestorianmoveoosaekuoft/nestorianmoveoosaekuoft_djvu.txt>
Franzmann, M., 'Yangzhou and Quanzhou: Ongoing Research on Syro-Turkic Inscriptions', in *From the Oxus River*, 83–92.
Franzmann, M. and S. Lieu, 'A New Nestorian Tombstone from Quanzhou: Epitaph of the Lady Kejamtâ', in *Jingjiao*, 293–302.
Gillman, I. and Hans-Joachim Klimkeit, 'Christian Monks and Hermits in Yuan China', in *Christians in Asia before 1500* (Ann Arbor, MI: University of Michigan Press, 1999), 295–8.
Halbertsma, T., *Early Christian Remains of Inner Mongolia: Discovery, Reconstruction and Appropriation* (Leiden: Brill, 2015).
Halbertsma, T., 'Some Field Notes and Images of Stone Sculpture Found at Nestorian Sites in Inner Mongolia', in *Hidden Treasures*, 51–69.

Hovorun, C., *From Antioch to Xi'an: an Evolution of "Nestorianism"* (Hong Kong: Chinese Orthodox Press, 2014; with a Chinese translation by Anna Chan).
Hunter, E., 'The Persian Contribution to Christianity in China: Reflections in the Xi'an Fu Syriac Inscriptions', in *Hidden Treasures*, 71–85.
Hunter, E., 'Commemorating the Saints at Turfan', in *Winds of Jingjiao*, 89–103.
Hunter, E. and J. F. Coakley, *A Syriac Service-Book from Turfan. Museum für Asiatische Kunst, Berlin MS MIK III 45* [Berliner Turfantexte XXXIX] (Turnhout: Brepols, 2017).
Hunter E. and M. Dickens, *Syrische Handschriften. 2, Syriac Manuscripts from the Berlin Turfan Collection* (Stuttgart, 2014).
Johnson, D., 'Syriac Crosses in Central and Southwest China', in *Winds of Jingjiao*, 41–61.
Keevak, M., *The Story of a Stele. China's Nestorian Monument and Its Reception in the West 1625–1916* (Hong Kong, 2008).
Latourette, K. S., *The Chinese, their History and Culture*, 2 vols (London: MacMillan, revised ed. 1964).
Lattimore, O., 'A Ruined Nestorian City in Inner Mongolia', *Geographical Journal* 84:6 (1934), 481–97.
Le Coq, A. von, *Chotscho* (Berlin: Dietrich Reimer, 1913).
Le Coq, A., *Buried Treasures of Chinese Turkestan: An Account of the Activities and Adventures of the Second and Third German Turfan Expeditions* (London: Allen and Unwin, 1928; repr. Routledge, 2018).
Legge, J., *The Nestorian Monument of Hsî-an Fû in Shen-hsî, China* (London, 1888/1924; repr. Piscataway, NJ: Gorgias Press, 2011).
Lieu, S., 'Nestorian Angels from Central Asia and Other Christian and Manichaean Remains at Zaitun (Quanzhou) on the South China Coast', in C. Benjamin and S. Lieu (eds), *Walls and Frontiers in Inner-Asian History*. Silk Road Studies 6 (Turnhout: Brepols, 2002), 1–17.
Lieu, S., 'Christian and Manichean Remains from Zaytun: An Introduction and Update', in I. Gardner et al. (eds), *From Palmyra to Zayton: Epigraphy and Iconography*. Silk Road Studies 10 (Turnhout, 2005), 189–206.
Lieu, S., 'Nestorian Remains of Zaytun (Quanzhou)', in *Jingjiao*, 277–92.
Lieu, S., 'The Romanitas of the Xi'an Inscription', in From the Oxus River, 123–40.
Lieu, S., 'Epigraphica Nestoriana Serica' <https://www.academia.edu/4535897/Epigr_Nest_Sinica>
Ligeti, L., 'Les sept monastères nestoriens de Mar Sargis', *Acta Orientalia Hungarica* 26: 2–3 (1972), 169–78.
Malek, R. and P. Hofrichter (eds), *Jingjiao: The Church of the East in China and Central Asia* (Sankt-Augustin: Institut Monumenta Serica, 2006).
Marsone, P., 'When Was the Temple of the Cross at Fangshan a "Christian Temple"?', in *Hidden Treasures*, 215–23.
Marsone, P., 'Two Portraits for One Man: George, King of the Önggüt', in From the Oxus River, 225–35.
Moule, A. C., *Christians in China before the Year 1550* (London, 1930; repr. Piscataway, NJ: Gorgias Press, 2011).

Nicolini, M., 'Eastern Outreach. The Monastic Mission to China in the Seventh to the Ninth Centuries', in *Mission and Monasticism, Studia Anselmiana* 158, *Analecta monastica* 13, (Rome, 2013), 63–70 <https://www.academia.edu/9950579/Eastern_Outreach._The_ Monastic_Mission_to_China_in_the_Seventh_to_the_Ninth_Centuries>

Nicolini-Zani, M., 'Past and Current Research on Tang Jingjiao Documents. A Survey', in *Jingjiao*, 23–44.

Nicolini-Zani, M., 'The Tang Christian Pillar from Luoyang and Its Jingjiao Inscription. A Preliminary Study', *Monumenta Serica* 57 (2009), 99–140.

Nicolini-Zani, M., 'Luminous Ministers of the Da Qin Monastery: A Study of the Christian Clergy Mentioned in the *Jingjiao* Pillar from Luoyang', in From the Oxus River, 141–60.

Nicolini-Zani, M., *Christian Monks on Chinese Soil. A History of Monastic Missions to China* (Collegeville, MN: Liturgial Press, 2016).

Niu, R., 'Nestorian Inscriptions from China (13th–14th Centuries)', in *Jingjiao*, 209–42.

Niu, R., 'A Comparative Study on the Nestorian Inscriptions from Semireche, Inner Mongolia and Quanzhou', in *Hidden Treasures*, 101–8.

Niu, R., *La croix-lotus: inscriptions et manuscrits nestoriens en écriture syriaque découverts en Chine (XIIIᵉ–XIVᵉ siècles)* (Shanghai, 2010).

Parry, K., 'Images in the Church of the East: The Evidence from Central Asia and China', *BJRL* 78:3 (1996), 143–62 <https://www.escholar.manchester.ac.uk/api/ datastream?publicationPid=uk-ac-man-scw:1m2404&datastreamId=POST-PEER- REVIEW-PUBLISHERS-DOCUMENT.PDF>

Parry, K., *From Palmyra to Zayton: Epigraphy and Iconography* (Turnhout: Brepols, 2005).

Parry, K., 'The Iconography of the Christian Tombstones from Zayton', in I. Gardner et al. (eds), *From Palmyra to Zayton: Epigraphy and Iconography*. Silk Road Studies 10 (Turnhout, 2005), 229–46.

Parry, K., 'The Art of the Church of the East in China', in *Jingjiao*, 321–39.

Pelliot, P., *Les grottes de Touen-Houang*, 2 vols (Paris, 1914, 1920).

Pelliot, P., 'Un témoignage éventuel sur le christianisme à Canton au XIᵉ siècle', *Mélanges chinois et bouddhiques* 1 (1931–2), 217–91.

Pelliot, P., *L'inscription nestorienne de Si-Ngan-Fou* (Paris, 1996).

Riboud, P., 'Le christianisme syriaque à l'époque Tang', in *Études syriaques* 12, 41–62.

Ruji, N., 'Nestorian Inscriptions from China (13th–14th Centuries)', in *Jingjiao*, 209–42.

Saeki, P. Y., *The Nestorian Documents and Relics in China* (Tokyo, 1937; repr. Piscataway, NJ: Gorgias Press, 2012).

See: The International Dunhuang Project: The Silk Road Online: <http://idp.bl.uk>

Sims-Williams, N., *The Christian Sogdian Manuscript C2*. Berliner Turfantexte 12 (Berlin: Akademie Verlag, 1985).

Sims-Williams, N., 'Bulayïq', in E. Yarshater (ed.), *Encyclopaedia Iranica*, vol. 4 (London: Routledge, 1990), 545.

Sims-Williams, N., 'Die christlich-sogdischen Handschriften aus Bulayiq' in *Ägypten, Vorder Asien, Turfan: Probleme des Edition* (Berlin: Akademie Verlag, 1991), 119–25; see also <http://www.iranicaonline.org/articles/bulayq-town-in-eastern-turkestan>

Sims-Williams, N., 'Sogdian and Turkish Christians in the Turfan and Tun-huang Manuscripts', in A. Cadonna (ed.), *Turfan and Tun-huang: The Texts. Encounter of Civilisations on the Silk Route*. Orientalia Venetiana 4 (Florence: Leo S. Olschki, 1992), 43–61.

Sims-Williams, N., 'A Greek-Sogdian Bilingual from Bulayïq', in *Convegno internazionale: La Persia e Bisanzio (Roma, 14–18 ottobre 2002), Atti del Convegno* (Rome: Accademia nazionale dei Lincei, 2004), 623–631.

Sims-Williams, N. (ed.), *Biblical and Other Christian Sogdian Texts from the Turfan Collection* (Turnhout: Brepols, 2014).

Smelova, N., 'Manuscrits chrétiens de *Qara Qoto*: nouvelles perspectives de recherche', Études syriaques 12, 215–36.

Standaert, N., *Handbook of Christianity in China*, Handbook of Oriental Studies, vol. 1, for the period 635–1800 (Leiden: Brill, 2001).

Takahashi, H., *Transcribed Names in Chinese Syriac Christian Documents* (Piscataway, NJ: Gorgias Press, 2009).

Takahashi, H., 'China, Syriac Christianity', in *Gorgias Dictionary*, 94–6.

Takahashi, H., 'Syriac Christianity in China', *Syriac World*, 625–52.

Tamcke, M., 'The Glorious Past: The History of the Church of the East in China as Symbol among East Syrian Christians in the 20th Century', in *From the Oxus River*, 457–68.

Tang, L., *A Study of the History of Nestorian Christianity in China and Its Literature in Chinese together with a New English Translation of the Dunhuang Nestorian Documents* (Frankfurt-am-Main: Peter Lang, 2002).

Tang, L., 'A Preliminary Study on the Jingjiao Inscription of Luoyang: Text Analysis, Commentary and English Translation', in *Hidden Treasures*, 109–32.

Tang, L., 'Medieval Sources on the Naiman Christians and on their Prince Küchlüg Khan', in *Hidden Treasures*, 257–66.

Tang, L., *East Syriac Christianity in Mongol-Yuan China (12th–14th Centuries)*. Orientalia Biblica et Christiana 18 (Wiesbaden: Harrassowitz Verlag, 2011).

Tang, L., 'A New Investigation into Several East Syrian ("Nestorian") Christian Epitaphs Unearthed in Quanzhou: Commentary and Translation', in D. Bumazhnov et al. (eds), *Bibel, Byzanz und Christlicher Orien* (Louvain: Peeters, 2011), 343–62.

Tang, L. 'Le christianisme syriaque en Chine dans la dynastie des Mongols Yuan: Diffusion, Statut des chrétiens et Déclin (XIIIe–XIVe siècles)', Études syriaques 12, 63–88.

Tang, L., 'Syriac Christianity in the Land of the Tangut (8th–14th Centuries)', in P. Edwell, Gunner Mikkelsen and K. Parry (eds), *Byzantium to China: Religious Encounters along the Silk Road. Festschrift for Samuel Lieu* (Leiden: Brill, 2020).

Tang, L. and D. Winkler, From the Oxus River to the Chinese Shores. Studies on East Syriac Christianity in China and Central Asia (Berlin: LIT Verlag, 2013).

Tang, L. and D. Winkler, Winds of Jingjiao. Studies on East Syriac Christianity in China and Central Asia (Vienna: LIT Verlag, 2016).

Thompson, G., 'Was Alopen a 'Missionary'?', in *Hidden Treasures*, 267–78.

Tisserand, E., 'Église nestorienne en Chine', *Dictionnaire de théologie catholique* 11.1 (1931), cols 199–207.

Twitchett, D. and J. Fairbank (eds), *The Cambridge History of China* (Cambridge: Cambridge University Press, 1978).

Wang D., 'Remnants of Christianity from Chinese Central Asia in Medieval Ages', in *Jingjiao*, 149–62.
Wang, Y., 'Priests of *Jingjiao* in the Xizhou Uighur Kingdom (Five Dynasties – The Early Song Period)', in *Winds of Jingjiao*, 333–46.
Wilmshurst, D., 'The Syrian Brilliant Teaching', *Journal of the Royal Asiatic Society, Hong Kong Branch* 30 (1990), 44–74 <http://hkjo.lib.hku.hk/archive/files/eba2efa1962585c63cfe8ff6183e7940.pdf>
Wilmshurst, D., 'Bet Sinaye: A Typical East Syrian Province?', in *Winds of Jingjiao*, 253–66.
Winkler, D. and L. Tang (eds), *Hidden Treasures and Intercultural Encounters: Studies on East Syriac Christianity in China and Central Asia* (Münster: LIT Verlag, 2009).
Wylie, A., *On the Nestorian Tablet of Se-Gan Foo* (Piscataway, NJ: Gorgias Press, 2013).
Zhang, G. and R. Xinjiang, 'A Concise History of the Turfan Oasis and Its Exploration', *Asia Major* 11 (1998), 13–37.
Zhang, X., 'Why Did Chinese Nestorians Name their Religion *Jingjiao* ?', in *Winds of Jingjiao*, 283–309.
Zieme, P., 'Turkic Christianity in the Black City (Xaraxto)', in From the Oxus River, 99–104.

Chapter 6: Under the Mongols (1206–1368) and Tamerlane (1370–1405)

Amar, J. P., 'Sawma, Rabban', in *Gorgias Dictionary*, 360–1.
Amar, J. P., 'Yahbalaha III', in *Gorgias Dictionary*, 429.
Amitai-Preiss, R., *Mongols and Mamluks: The Mamluk-Īlkhānid War, 1260–1281* (Cambridge: Cambridge University Press, 1995).
Amitai-Preiss, R. (ed.), *The Mongol Empire and Its Legacy* (Leiden: Brill, 1999).
Amitai-Preiss, R., *The Mongols in the Islamic Lands: Studies in the History of the Ilkhanate* (Aldershot: Ashgate, 2007).
Atwood, C., *The Encyclopedia of Mongolia and the Mongol Empire* (New York: Facts on File, 2004).
Baum, W., 'The Age of the Mongols Thirteenth and Fourteenth Centuries', in *A Concise History*.
Baumer, C., *The History of Central Asia. The Age of Islam and the Mongols*, 3 vols (London: Routledge, 2016).
Bedjan, P., *Histoire de Mar Jaballaha, patriarche et Raban Sauma* (Paris: Maisonneuve, 1888).
Biran, M. and K. Hodong (eds), *The Cambridge History of the Mongol Empire*, 2 vols (Cambridge: Cambridge University Press, 2015).
Borbone, P. G., *Un ambassadeur du Khan Argun en Occident, Histoire de Mar Yahlallaha III et de Rabban Sauma (1281–1317)* (Paris: L'Harmattan, 2008).
Boyle, J., *The Successors of Genghis Khan* (New York: Columbia University Press, 1971).

Boyle, J. (trans.), *Genghis Khan: The History of the World-Conqueror* (Seattle: University of Washington Press, 1997).
Budge, W. E. A., *The Monks of Kublai Khan Emperor of China. The History of the Life and Travels of Rabban Sawma and Mar Yahballaha III* (London, 1928; repr. Piscataway, NJ: Gorgias Press, 2017).
Chabot, J. B. (ed. and trans.), *Histoire du Patriarche Mar Jabalaha III et du moine Rabban Cauma*, Revue de l'Orient Latin 1 (1893), 566–610; 2 (1894), 73–143, 235–300.
Chaliand, G., *Les empires nomades: de la Mongolie au Danube: V^e–IVe siècles av. J.-C. – XVe -XVIe siècles ap. J.-C.* (Paris: Perrin, 2005).
Dauvillier, J., 'Guillaume de Rubrouck et les communautés chaldéennes d'Asie centrale au Moyen-Âge', *Annuaire de l'école des législations religieuses* 2 (1951–2), 36–42.
Favereau, M. (ed.), *La Horde d'Or et l'islamisation des steppes eurasiatiques / The Golden Horde and the Islamisation of the Eurasian Steppes* (2018) <https://journals.openedition.org/remmm/10234> (see the maps in the introduction by M. Favereau).
Favereau, M., 'The Islamisation of the Steppe: Introduction' <https://www.academia.edu/39871995/The_Islamisation_of_the_Steppe_Introduction?auto=download>
Fiey, J. M., *Chrétiens syriaques sous les Mongols (Il-Khanat de Perse, XIIIe–XIVe s)*, CSCO 362/44 (Louvain: Peeters, 1975).
Fiey, J. M., 'Le grand catholicos turco-mongol Yahwalaha III (1281–1317)', *Proche-Orient chrétien* 38:3 (1988), 209–28.
Golden, P., 'Religion among the Qipcaqs of Medieval Eurasia', *Central Asiatic Journal* 42 (1998), 180–237.
Grousset, R., *L'Empire des steppes: Attila, Gengis-Khan, Tamerlan* (Paris: Payot, 2001).
Juvaini, A. M., *Genghis Khan: The History of the World-Conqueror*, trans. J. A. Boyle (Manchester, 1958).
Marozzi, J., *Tamerlane: Sword of Islam, Conqueror of the World* (London: Harper Collins, 2004).
Montgomery, J. A. (trans.) *The History of Yaballaha III and His Vicar Bar Sauma* (New York, 1927; repr. Piscataway, NJ: Gorgias Press 2006).
Murre van den Berg, H., 'The Church of the East in Mesopotamia in the Mongol Period', in *Jingjiao*, 377–94.
Ostrowski, D., Muscovy and the Mongols: Cross-Cultural Influences on the Steppe Frontier, 1304–1589 (Cambridge: Cambridge University Press, 2010).
Pelliot, M., 'Les Mongols et la papauté', *Revue de l'Orient chrétien* 23 (1922–3), 3–30.
Rossabi, M., 'Khubilai Khan and the Women in his Family', in W. Bauer (ed.), *Studia Sino-Mongolica: Festschrift für Herbert Franke* (Wiesbaden, 1979), 158–66.
Roux, J.-P., *La Naissance du monde chez les Turcs et les Mongols* (Paris, 1959).
Roux, J.-P., *La religion des Turcs et des Mongols* (Paris: Payot, 1984).
Roux, J.-P., *Histoire de l'Empire mongol* (Paris: Fayard, 1993; repr. 2008).
Roux, J.-P., *Gengis Khan et l'empire mongol* (Paris: Gallimard, 2002).
Tang, L., 'Sorkaktani Beki: A Prominent Nestorian Woman at the Mongol Court', in *Jingjiao*, 349–56.
Tang, L., *East Syriac Christianity in Mongol-Yuan China* (Wiesbaden, 2011).

Tang, L., 'Le Christianisme syriaque dans la Chine des Mongols Yuan: diffusion, statut des chrétiens et déclin (XIIIe–XIVe siècles)', in *Études syriaques* 12, 63–88.
Tang, T., 'Rediscovering the Ongut King George: Remarks on a Newly Excavated Archaeological Site', in *From the Oxus River*, 255–66.
Tisserand, E., 'Les Nestoriens sous les Mongols', *Dictionnaire de théologie catholique* 11:1 (Paris: Letouzey et Ané, 1931), col. 213–18.
Weatherford, J., *Genghis Khan and the Making of the Modern World* (New York: Three Rivers Press, 2005); see chapter 8, 'Khubilai Khan and the New Mongol Empire', 195–217.

Travellers

Jackson, P. (trans.), *The Mission of Friar William of Rubruck: His Journey to the Court of the Great Khan Möngke (1253–1255)* (London: Routledge, 1990).
Latham, R. (trans.), *The Travels of Marco Polo* (Harmondsworth: Penguin Books, 1958).
Polo, M., *Le devisement du monde*, eds A. Moule and P. Pelliot (Paris: La Découverte, 2011).
For contemporary account of travels in Central Asia and China, see the books of travellers and explorers, for example, those of Peter Fleming and Ella Maillart (on horseback).

Chapter 7: The Nineteenth Century

Aprem (Mar), *Western Missions Among Assyrians* (Thrissur, 1982).
Ardalan, S., *Les Kurdes Ardalân entre la Perse et l'Empire ottoman* (Paris, 2004).
Arfa, H., *The Kurds. An Historical and Political Study* (Oxford: Oxford University Press, 1966).
Binder, H., *Au Kurdistan, en Mésopotamie et en Perse* (Mission scientifique du ministère de l'Instruction publique) (Paris, 1887).
Bois, T., 'Kurdes et Kurdistan', *Encyclopédie de l'Islam* 5 (1986), 421–89.
Campanile, G., *Histoire du Kurdistan* (1818; repr. Paris: Harmattan, 2004).
Chevalier, M., *Les montagnards chrétiens du Hakkâri et du Kurdistan septentrional* (Paris: Département de Géographie de l'université de Paris-Sorbonne, 1985).
Coakley, J. F., *The Church of the East and the Church of England: A History of the Archbishop of Canterbury's Assyrian Mission* (Oxford: Clarendon Press, 1992).
Coakley, J. F., 'A List of Assyrian Villages in Persia, August 1893', *Journal of the Assyrian Academic Society* 7:2 (1993), 41–55.
Coan, F., *Missionary Life in the Middle East, or, Yesterday in Persia and Kurdistan* (Claremont, 1939; repr. Piscataway, NJ: Gorgias Press, 2006).
Fiey, J. M., 'Proto-histoire chrétienne du Hakkari turc', *L'Orient Syrien* 9 (1964), 443–72.
Fiey, J. M., 'Hakkari', *Dictionnaire d'histoire et de géographie ecclésiastiques* 23 (1990), 120–1.

Heazell, F. (ed.), *Kurds and Christians* (London: Wells, Gardner & Co., 1913; repr. Piscataway, NJ: Gorgias Press 2004).
Hornus, J. M., 'Rapport du consul de France à Erzéroum sur la situation des chrétiens en Perse au milieu de XIXᵉ siècle', *Proche Orient Chrétien* 21 (1971).
Joseph, J., *The Nestorians and Their Muslim Neighbours: A Study of Western Influence on their Relations* (Princeton: Princeton University Press, 1961).
Joseph, J., *Muslim-Christian Relations and Inter-Christian Rivalries in the Middle East* (New York: SUNY Press, 1983).
Joseph, J., *The Modern Assyrians of the Middle East. Encounters with Western Christian Missions, Archaeologists and Colonial Powers* (Leiden: Brill, 2000).
Laurens, H., *Les crises d'Orient*, vol. 1, *Question d'Orient et Grand Jeu (1768–1914)* (Paris: le Grand Livre du mois, 2017).
Nikitine, B., 'Le système routier du Kurdistan (Le pays entre les deux Zab)', *Géographie* 58 (June 1935), 363–85.
'Patriarchal church of the Church of the East Hakkari' <http://www.jelleverheij.net/monuments/patriarchal-church-of-the-church-of-the-east.html>
Rondot, P., 'Les tribus montagnardes de l'Asie antérieure, quelques aspects sociaux des populations kurdes et assyriennes', *Bulletin d'Études orientales de l'Institut français de Damas* 6 (1937), 1–50.
Rondot, P., 'Origine et caractère ancestraux du peuplement assyrien en Haute Djézireh syrienne. Esquisses d'une étude de la vie tribale', *Bulletin d'Études orientales de l'Institut français de Damas* 41–2 (1989–90), Damascus: Institut français (1993), 65–111.
Sado, S., 'La mission orthodoxe russe à Ourmiah (1898–1918)', *Christian Reading* 13, Theological Academy of St. Petersburg (1996), 73–112. (in Russian).
Tisserand, E., 'Église nestorienne, Époque moderne', *Dictionnaire de théologie catholique* 11:1 (1931), cols 255–60.
Vine, A. R., *The Nestorian Churches: A Concise History of Nestorian Christianity in Asia from the Persian Schism to the Modern Assyrians* (London: Independent Press, 1937).
Wilmshurst, D., 'The Age of the European Missions', in D. Wilmshurst, *The Martyred Church* (2011), 368–94.

Missionaries, Travellers and Archaeologists

Ainsworth, J. W., *Travels and Researches in Asia Minor, Mesopotamia, Chaldea and Armenia* (London: J. W. Parker, 1842).
Armajani, Y., 'Christian Missions in Persia', *Encyclopaedia iranica* 5:2 (1992), 544–7.
Badger, G. P., *The Nestorians and Their Rituals: With the Narrative of a Mission to Mesopotamia and Coordistan in 1842–1844 and of a Late Visit to Those Countries in 1850* (London: Joseph Masters, 1852; repr. Farnborough: Gregg International, 1969).
Bird Bishop, I., *Journeys in Persia and Kurdistan* (London: John Murray, 1891).
Coan, F., *Missionary Life in the Middle East, or, Yesterday in Persia and Kurdistan* (Claremont, 1939; repr. Piscataway, NJ: Gorgias Press, 2006).
Cutts, E. L., *The Assyrian Christians: Report of a Journey to the Christians in Koordistan and Oroomiah* (London, 1877; repr. London: Forgotten Books, 2019).

Fletcher, J. P., *Notes from Nineveh and Travels in Mesopotamia, Assyria, and Syria* (London, 1850) < http://www.aina.org/books/nfn/nfn.htm>

Grant, A., *The Nestorians or the Lost Tribes* (London, 1841; repr. Piscataway, NJ: Gorgias Press, 2004).

Laurie, T., *Dr Grant and the Mountain Nestorians* (Boston, 1856; repr. Gorgias Press, 2005).

Layard, A. H., *Discoveries in the Ruins of Nineveh and Babylon: With Travels in Armenia, Kurdistan and the Desert: Being the Result of a Second Expedition Undertaken for the Trustees of the British Museum* (London, 1853) <http://www.aina.org/books/dan.htm> (see Chapter 7).

Layard, A. H., *Nineveh and Its Remains: With an Account of* a *Visit to the Chaldean Christians of Kurdistan, and the Yezidis* (New York and London: Appleton and Company, 1854).

Maclean A. J. and W. H. Browne, *The Catholicos of the East and His People Being the Impression of Five Years' Work in the 'Archbishop of Canterbury's Assyrian Mission'* (London, 1892; repr. Piscataway, NJ: Gorgias Press, 2006).

Perkins, J., *A Residence of Eight Years in Persia among the Nestorian Christians* (Andover, 1843; repr. as *Missionary Life in Persia*. Piscataway, NJ: Gorgias Press, 2006).

Riley, J. A., *Report on the Foundation of the Archbishop's Mission to the Assyrian Church in 1886* (London, 1887).

Riley, J. A., *The Archbishop of Canterbury's Mission to the Assyrian Christians* (London, 1891).

Shedd, M. L., *The Measure of a Man: The Life of William Ambrose* Shedd, *Missionary to Persia* (repr. Piscataway, NJ: Gorgias Press, 2006).

Wigram, W. A., *An Introduction to the History of the Assyrian Church, or, The Church of the Sassanid Persian Empire, 100–640 A.D., 1910 (repr. Picataway, NJ: Gorgias Press, 2004)*.

Wigram, W. A., The Cradle of Mankind. Life in Eastern Kurdistan (London, 1914) <http://www.aina.org/books/com/com.htm> (see photos).

Wigram, W. A., *The Assyrians and Their Neighbours* (London: G. Bell & Sons, 1929; repr. Piscataway, NJ: Gorgias Press, 2002).

Chapter 8: The Twentieth Century

Aboona, H., *Assyrians, Kurds, and Ottomans: Intercommunal Relations on the Periphery of the Ottoman Empire* (New York: Cambria Press, 2008).

Aprem (Mar), *The Assyrian Church of the East in the Twentieth Century* (Kottayam: St Ephrem Ecumenical Research Institute, 2003).

Aprem (Mar), *Patriarch Mar Dinkha IV:* The Man and His Message (Trichur: Mar Narsai Press, 2004).

Austin, H., *The Baqubah Refugee Camp. An Account of Work on Behalf of the Persecuted Assyrian Christians* (London: Faith Press, 1920; repr. Gorgias Press, 2006).

Bozarslan, H., *Histoire de la Turquie, de l'Empire à nos jours* (Paris: Tallandier, 2015).

Coakley J. F., 'The Church of the East Since 1914', *BJRL* 78 (1996), 179–98.

Coakley, S. and J. Coakley, 'Church of the East', in *Gorgias Dictionary*, 99–100.

Caujole, P., *Les tribulations d'une ambulance française en Perse* (Paris: Les Gémeaux, 1921).
Chirol, V., *The Middle Eastern Question* (London, 1903; repr. 1930).
Dadesho, S., *The Assyrian National Question at the United Nations (A Historical Injustice Redressed)* (Modesto, CA, 1987).
Dauphin, C., 'The Rediscovery of the Nestorian Churches of the Hakkari (South Eastern Turkey)', *Eastern Church Review* 8 (1976), 56–67.
Deniz, F., 'Maintenance and Transformation of Ethnic Identity: The Assyrian Case', in O. Cetrez, S. Donabed and A. Makko (eds), *The Assyrian Heritage. Threads of Continuity and Influence* (Philadelphia, PA: Coronet Books, 2012), 319–31.
Donabed, S. and S. Mako, 'Between Denial and Existence: Situating Assyrians within the Discourse on Cultural Genocide', in O. Cetrez, S. Donabed and A. Makko (eds), *The Assyrian Heritage. Threads of Continuity and Influence* (Philadelphia, PA: Coronet Books, 2012), 281–95.
Donabed, S. G., *Reforging a Forgotten History: Iraq and Assyrians in the Twentieth Century* (Edinburgh: Edinburgh University Press, 2015).
Eshai Shimun (Mar), Catholicos Patriarch, *The Assyrian Tragedy* (Annemasse, 1934; repr. Bloomington, IN: Xlibris Corporation, 2010) <http://www.learnassyrian.com/assyrianlibrary/assyrianbooks/History%20-%201800AD%20-%20Present/Assyrian%20Tradegy%20-%20Mar%20Eshai%20Shimun%20XXIII.pdf>
Gaunt, D., *Massacres, Resistance, Protectors: Muslim-Christian Relations in Eastern Anatolia During World War I* (Piscataway, NJ: Gorgias Press, 2006).
Gaunt, D., 'The Assyrian Genocide of 1915' (Assyrian Genocide and Research Center, Los Angeles, CA, 2009) <http://www.seyfocenter.com/english/38/>
Gaunt, D., 'The Complexity of the Assyrian Genocide', in *Le génocide des Arméniens* (Paris: Conseil scientifique international pour l'étude du génocide des Arméniens, 2015), 70–91.
Gaunt, D. et al., *Let Them Not Return: Sayfo – The Genocide of the Assyrian, Syriac and Chaldean Christians in the Ottoman Empire* (New York: Berghahn Books, 2017).
Hellot-Bellier, F., *Chroniques de massacres annoncés. Les Assyro-Chaldéens d'Iran et du Hakkari face aux ambitions des empires 1896–1920* (Paris: Geuthner, 2014).
Holquist, P., 'Forms of Violence during the Russian Occupation of Ottoman Territory and in Northern Persia (Urmia and Astrabad)', in O. Bartov and E. Weitz (eds), *Shatterzone of Empires* (Indianapolis, IN: Indiana University Press, 2013), 317–61.
Holquist, P., 'The World Turned Upside Down: Refugee Crisis and Military Massacres in Occupied Northern Persia (1917–18)', in *Le génocide des Arméniens* (Paris: Conseil scientifique international pour l'étude du génocide des Arméniens, 2015), 130–54 (see the maps).
Husry, K., 'The Assyrian Affair of 1933', *International Journal of Middle East Studies* 5 (1974), Part I, 161–76; Part II, 344–60.
Internet Site about Patriarche Shimoun XXI Eshai : <http://marshimun.com>
Laurens, H., *Les crises d'Orient*, vol. II *La Naissance du Moyen-Orient (1919–1949)* (Paris: le Grand Livre du mois, 2019).
League of Nations, 'Civil, Religious and Political Status of the Assyrians', in *Settlement of the Assyrians of Iraq, Report of the Committee of the Council on the Settlement of the Assyrians of Iraq in the Region of the Ghab (French Mandated Territories of the Levant)*, 1935 <http://www.aina.org/books/sota.htm>

Lilian, M. Y., *Assyrians of the Van District during the Rule of Ottoman Turks*, 1914 <http://www.aina.org/books/aov.htm>
Makiya, K., *Republic of Fear: The Politics of Modern Iraq* (Oakland, CA: University of California Press, 1998).
Makko, A., 'Discourse, Identity and Politics: A Transnational Approach to Assyrian Identity in the Twentieth Century', in O. Cetrez, S. Donabed and A. Makko (eds), *The Assyrian Heritage. Threads of Continuity and Influence* (Philadelphia, PA: Coronet Books, 2012), 297–311.
Malek, Y., *The British Betrayal of the Assyrians* (Chicago: Joint-Action of the Assyrian National Federation & the Assyrian National League of America, 1935) <http://www.aina.org/books/bbota.htm>
McCarthy, J., *The Ottoman Turks: An Introductory History to 1923* (New York: Routledge, 1997).
Murre van den Berg, H., 'Syriac Identity in the Modern Era', in *Syriac World*, 770–82.
Naby, E., *The Assyrian Experience: Sources for the Study of the 19th and 20th Centuries: From the Holdings of the Harvard University Libraries* (with a selected bibliography) (Cambridge, MA: Harvard College Library, 1999).
Nikitine, B., 'Une petite nation victime de la guerre', *Revue des sciences politiques* 44:4 (1921), 602–24.
Nikitine, B., 'Les Kurdes et le christianisme', *Revue de l'Histoire des Religions* 85 (1922), 1–10.
Nikitine, B., 'La vie domestique des Assyro-Chaldéens du Plateau d'Ourmiah', Extrait de *L'Ethnographie* 11–12 (1925), 356–80.
Nikitine, B., *La Perse que j'ai connue* (Paris, 1941).
Omissi, D., 'Britain, the Assyrians and the Iraq Levies, 1919–1932', *Journal of Imperial and Commonwealth History* 17:3 (1989), 301–22.
Palmer, A., *The Decline and Fall of the Ottoman Empire* (New York, 1992).
Rogan, E., *The Fall of the Ottomans* (London amd New York: Faber & Faber, 2015).
Royel, Mar Awa. Biography of His Holiness Mar Gewargis III, Holy Catholic Apostolic Assyrian Church of the East Official News Website (Assyrian Church of the East, 2015) <https://en.wikipedia.org/wiki/Gewargis_III>
Sanders, J. C., *Assyrian-Chaldean Christians in Eastern Turkey and Iran. Their Last Homeland Re-Chartered* (Nijmegen: Kasteel Hernen, 1999) (see the maps).
Sarafian, A., *The Treatment of Armenians in the Ottoman Empire, 1915–16: Documents Presented to Viscount Grey of Fallodon by Viscount Bryce* (Reading: Taderon Press, 2005).
Schayegh, C. and A. Arsan (eds), *The Routledge Handbook of the Middle East Mandates* (London: Routledge, 2015).
Shahbaz, Y., *The Rage of Islam. An Account of the Massacre of Christians by the Turks in Persia* (Philadelphia, 1918; repr. Piscataway, NJ: Gorgias Press, 2006).
Shimmon, P., Massacres of Syrian Christians in North-West Persia and Kurdistan (New York: Columbia University Press, 1916).
Solhkhah, N., *The Assyrian Martyr Mar Benjamin Shimun Patriarch of the Church of the East* (Chicago, 2008) <http://www.learnassyrian.com/assyrianlibrary/assyrianbooks/Religion/Mar%20Benjamin%20Shimon%20-%20The%20Assyrian%20Martyr%20-%20Dr.%20Norman%20Solhkhah.pdf>
Stafford, R. S., 'Iraq and the Problem of the Assyrians', *International Affairs* 13:2 (1934), 159–85.
Stafford, R., *The Tragedy of the Assyrians* (1935; repr. Piscataway, NJ: Gorgias Press, 2006) <http://www.aina.org/books/tota.htm>

Stavridis, S., *Lady Surma: The Pillar of the Assyrian Nation 1883–1975* <https://www.academia.edu/8445717/Lady_Surma_the_pillar_of_the_Assyrian_nation_1883-1975>

Surma d'Bait Mar Shimun, *Assyrian Church Customs and the Murder of Mar Shimun* (London, 1920; repr. New York: Vehicle Editions, 1983).

Talay, S. and S. Barthoma, *Sayfo 1915, an Anthology on the Genocide of Assyrians/Arameans During the First World War* (Piscataway, NJ: Gorgias Press, 2018).

'The 1933 Massacre of Assyrians in Simmele, Iraq', <http://www.aina.org/releases/20040805022140.htm>

Travis, H., The Assyrian Genocide: Cultural and Political Legacies. Routledge Studies in Modern History (London: Routledge, 2017).

Weibel Yacoub, C., *Surma l'Assyro-Chaldéenne (1883–1975). Dans la tourmente de Mésopotamie* (Paris: Harmattan, 2007).

Weibel Yacoub, C., *Le rêve brisé des Assyro-Chaldéens. L'introuvable autonomie* (Paris: Éd. du Cerf, 2011).

Weibel Yacoub, C., *La France et les Assyro-Chaldéens. Qu'en dit la presse?* (Paris: Harmattan, 2019).

Werda, J., *The Flickering Light of Asia or the Assyrian Nation and Church* (Jersey City: self-published, 1924) <http://www.aina.org/books/fla/fla.pdf>

Wigram W. A., *Our Smallest Ally: A Brief Account of the Assyrian Nation in the Great War* (London: SPCK, 1920).

Winkler, D., 'The Twentieth Century', in *A Concise History*, 135–57.

Winkler, D., 'East Syriac Christianity in Iraq: A Glance at History from the First World War until Today', in *Hidden Treasures*, 321–34.

Yacoub, J., *The Assyrian Question* (Chicago: Alpha Graphic, 1986).

Yacoub, J., *Qui s'en souviendra ? 1915: le génocide assyro-chaldéo-syriaque* (Paris: Éd. du Cerf, 2014).

Yacoub, J., *Year of the Sword. The Assyrian Christian Genocide: A History*; trans. J. Ferguson (London: Hurst & Company, 2016).

Yacoub, J., *Une diversité menacée, les Chrétiens d'Orient face au nationalisme et à l'islamisme* (Paris: Salvator, 2018).

Yohannan, A., The Death of a Nation; or the Ever Persecuted Nestorians; or, Assyrian Christians (New York and London: G. P. Putnam's Sons, 1916; repr. 2009).

Yonan, G., *Lest we Perish: A Forgotten Holocaust* (1996 (Books online <www.aina.org>)).

Zubaida, S., 'Contested Nations: Iraq and the Assyrians', *Nations and Nationalism 6:3 (2000)*, 363–82.

The Assyrian Genocide

Khosroeva, A., 'The Assyrian Genocide in the Ottoman Empire and Adjacent Territories', in R. Hovannisian (ed.), *The Armenian Genocide: Cultural and Ethical Legacies* (New Brunswick: Transaction Publishers, 2007), 267–274.

Khosroeva, A., *The Assyrian Genocide during World War I* (Chicago: Center for Assyrian Genocide Studies, 2012).

Khosroeva, A., 'The Ottoman Genocide of the Assyrians in Persia', in H. Travis (ed.), *The Assyrian Genocide: Cultural and Political Legacies* (London and New York: Routledge, 2017), 137–57.
Khosroeva, A., 'Assyrians in the Ottoman Empire and the Official Turkish Policy of their Extermination, 1890s–1918', in G. Shirinian (ed.), *Genocide in the Ottoman Empire: Armenians, Assyrians and Greeks 1913–1923* (New York, NY: Berghahn Books, 2017), 105–31.
Khosroeva, A., 'The Significance of the Assyrian Genocide after a Century', in S. Talay and S. Barthoma (eds), *Sayfo 1915: An Anthology on the Genocide of Assyrians/Arameans During the First World War* (Piscataway, NJ: Gorgias Press, 2018), 61–9.
Naayem, J., *Les Assyro-Chaldéens et les Arméniens massacrés par les Turcs* (Paris, 1920, repr. Paris: Le Cercle d'écrits caucasiens, 2016).

The Armenian Genocide

Akçam, T., *The Young Turks' Crime against Humanity: The Armenian Genocide and Ethnic Cleansing in the Ottoman Empire* (Princeton, NJ: Princeton University Press, 2012).
Dadrian, V., *The History of the Armenian Genocide: Ethnic Conflict from the Balkans to Anatolia to the Caucasus* (New York: Berghahn Books, 2004).
Kévorkian, R., *Le génocide des Arméniens* (Paris: O. Jacob, 2006).
Le génocide des Arméniens: cent ans de recherche, 1915–2015: actes du colloque international, Paris, Sorbonne, organisé par le Conseil scientifique international pour l'étude du génocide des Arméniens. Mémorial de la Shoah, École des hautes études en sciences sociales (Paris: Armand Colin, 2015).
Lepsius, J., *Rapport secret sur les massacres d'Arménie* (Paris: Payot, 1919).
Mandelstam, A., *Le sort de l'Empire ottoman* (Lausanne: Payot, 1917).

Chapter 9: The Twenty-First Century around the World. The Diaspora

AINA News, « A Dialogue on the Assyrian Homeland and Diaspora » <http://www.aina.org/news/20190618170330.htm>
Aro, S. and R. M. Whiting (eds), *The Heirs of Assyria: Proceedings of the Opening Symposium of the Assyrian and Babylonian Intellectual Heritage Project Held in Tvärminne, Finland, 1998* (University Park, PA: Penn State University Press, 2000).
Baumer, C., 'Recent Archaeological Discoveries and Ecclesiastical and Political Developments', in Baumer 2016, 287–93.
Brock, S., 'Diaspora', in *Gorgias Dictionary*, 119–22.
Consulter <http://www.aina.org>
Frahm, E. (ed), A Companion to *Assyria* (New Haven, CN: Yale University Press, 2017).
Goldschmidt, A., *A Concise History of the Middle East*, 12th ed. (London: Routledge, 2018).

Heyberger, B., *Les chrétiens d'Orient*. Que-sais-je? (Paris: Presses universitaires de France, 2017).
Holy Apostolic Catholic Assyrian Church of the East Official News Website <https://news.assyrianchurch.org/>
Holy Synod in May 2019 (Concerning Dioceses and the Election of Bishops of the Church of the East) <https://news.assyrianchurch.org/wp-content/uploads/2019/06/2019-General-English-Synod-Decrees-Public.pdf?sfns=mo>
Hunter, E., 'Changing demography: Christians in Iraq since 1991', in *Syriac World*, 783–96.
Isakhan, B. et al., 'Cultural Cleansing and Iconoclasm under the Islamic State: Attacks on Yezidis and Christians and their Heritage', in F. Oruc (ed.), *Sites of Pluralism: Community Politics in the Middle East* (London: Hurst & Company, 2019), 181–94.
Kenneth, R., M. Tadros and T. Johnson (eds), *Christianity in North Africa and West Asia* (Edinburgh: Edinburgh University Press, 2018).
Naby, E., *The Assyrians of the Middle East: The History and Culture of a Minority Christian Community* (London: I. B. Tauris, 2019).
Parpola, S., 'National and Ethnic Identity in the Neo-Assyrian Empire and Assyrian Identity in Post-Empire Times', *Journal of Assyrian Academic Studies* 18:2 (2004), 5–22 < http://jaas.org/edocs/v18n2/Parpola-identity_Article%20-Final.pdf>
Tamcke, M., 'The Emigration of Syriac Christians from the Middle East: Motives and Impact', in D. Winkler (ed.), Syriac Christianity in the Middle East and India: Contributions and Challenges. Pro Oriente Studies in Syriac Tradition 1 (Piscataway, NJ: Gorgias Press, 2013).
Yacoub, J., 'La diaspora assyro-chaldéenne', *L'Espace géographique* 1 (1994), 29–37.
Yacoub, Y., *Babylone chrétienne. Géopolitique de l'Église de Mésopotamie* (Paris: Desclée de Brouwer, 1996).
Yacoub, J., *Une diversité menacée. Les Chrétiens d'Orient face au nationalisme arabe et à l'islamisme* (Paris : Salvator, 2018).
Yonan, G., *Assyrer heute* (Hamburg: Gesellschaft für bedrohte Völker, 1978).

Middle East

Iraq

Benraad, M., *Irak, La revanche de l'histoire. De l'occupation étrangère à l'État islamique* (Paris: Vendémiaire, 2015).
Chaillot, C., 'L'Église assyrienne apostolique en Iraq', *Proche-Orient chrétien* 67:1–2 (2017), 62–74.
Fiey, J. M., 'Sanctuaires et villages syriaques orientaux de la Vallée de la Sapna', *Le Muséon* 102 (1989), 43–67.
Hunter, E. (ed.), *The Christian Heritage of Iraq* (Piscataway, NJ: Gorgias Press, 2009).
Naby, E., 'The Plight of Christians in Iraq', *The New York Review of Books*, volume 53, no 19, 2006.
Petrosian, V., 'Assyrians in Iraq', in *Iran and the Caucasus* 10:1 (2006), 113–47.

Talay, S., 'Die Christen im Nord Irak – Sprache und Namenbezeichnung', in T. Peral and H. Oberkampf (eds), *Heimat oder Exil ? Zur Lage der Christen im Irak* (Neuendettelsau: Erlanger Verlag für Mission und Ökumene, 2013), 184–95.

Taneja, P., *Assimilation, Exodus, Eradication: Iraq's Minority Communities since 2003* (London: Minority Rights Groups International, 2007).

Teule, H., 'Iraq', in K. Ross, M. Tadroz and T. Johnson (eds), *Christianity in North Africa and West Asia* (NAWA) (Edinburgh: Edinburgh University Press, 2018), 164–76.

Yacoub, J., *Menaces sur les chrétiens d'Irak* (Paris: Chambray-lès-Tours, 2003).

Youkhana, E., 'Assyrer und Kurden – Schwierige Geschichte und Gemeinsame Zukunft ?', in T. Peral and H. Oberkampf (eds), *Heimat oder Exil? Zur Lage der Christen im Irak* (Neuendettelsau: Erlanger Verlag für Mission und Ökumene, 2013), 196–207.

Iran

Duval, R., 'Inscriptions de Salamas en Perse', *Journal asiatique* 1 (1885), 39–62 (repr. Piscataway, NJ: Gorgias Press, 2011).

Fiey, J. M., 'Adarbaygan chrétien', *Le Muséon* 86 (1973), 397–435.

Hellot-Bellier, F., 'L'apport des inscriptions syriaques à la connaissance des chrétiens d'Ourmia', in Briquel 2004, 117–23.

Hellot-Bellier, F., 'Églises de l'Azerbaïdjan iranien et du Hakkari', in Briquel 2013, 421–40.

Ishaya, A., 'From Contributions to Diaspora: Assyrians in the History of Urmia, Iran', *JAAS* 16:1 (2002), 55–76 <http://www.nineveh.com/Assyrians%20in%20the%20History%20of%20Urmia,%20Iran.html>

Jeloo, N., 'Evidence in Stone and Wood: The Assyrian/Syriac History and Heritage of the Urmia Region in Iran, as Reconstructed from Epigraphic Evidence', *Paroles de l'Orient* 35 (2010), 1–15.

Jeloo, N., 'The Socio-Cultural History and Heritage of Urmia's Ethnic Assyrians, Based on a Corpus of Syriac and Neo-Aramaic Inscriptions' (doctoral dissertatiom, University of Sydney, 2013).

Jeloo, N., 'Persian Christians: Assyrian Art and Architecture of Urmia', *Paroles de l'Orient*, 40 (2015), 1–13.

Macuch, R., 'Assyrians in Iran I: The Assyrian community (Āšūrīān) in Iran', Encyclopaedia Iranica 2 (1987), 817–24 <http://www.iranicaonline.org/articles/assyrians-in-iran-i-community>

Naby, E., 'The Assyrians of Iran: Reunification of a 'Millat', 1906–1914', *International Journal of Middle East Studies* 8:2 (1977), 237–49.

Naby, E., 'Ishtar: Documenting the Crisis in the Assyrian Iranian Community', *Middle East Review of International Affairs* 10:4 (December 2006), 92–102.

Syria

Bohas, G., *Les Araméens du bout du monde* (Toulouse: Éd. universitaires du Sud, 1994).

Chaillot, C., 'Proche-Orient: le destin incertain des Assyriens, chrétiens oubliés entre État Islamique et nationalisme kurde', *Religioscope* (31 Oct 2015) <http://religion.info/french/entretiens/article_670.shtml>

Fernandez, A. M., 'Dawn at Tell Tamir: The Assyrian Christians on the Khabur River', *Journal of Assyrian Academic Studies* 12:1 (1998), 34–46.

ISIS Execute Three Assyrians in Syria <http://www.aina.org/news/20151008022445.htm>

ISIS Release 22 Assyrian Hostages in Syria <http://www.aina.org/news/20160129142745.htm>

Oehring, O., *The Situation and Prospects of Christians in North and North-East Syria*, 2019 <https://www.kas.de/documents/252038/4521287/The+Situation+and+Prospects+of+Christians+in+North+and+North-East+Syria.pdf/b2a2e1b4-2427-10d4-a9a2-d0eb2bd578e9?version=1.0&t=1569924961335>

North America

United States and Canada

Aprem, Mar, *Israel, Scandinavia and USA* (Thrissur, 1996).

Ishaya, A., 'Class and Ethnicity in Rural California: The Assyrian Community of Modesto-Turlock, 1910–1985' (dissertation of the University of California at Los Angeles, 1985).

Ishaya, A., 'Assyrians-Americans: A Study in Ethnic Reconstruction and Dissolution in Diaspora', *JAAS* 17:2 (2003), 19–38.

Ishaya, A., 'Settling in Diaspora: A History of Urmia Assyrians in the United States', *JAAS* 20:1 (2006), 4–27.

Ishaya, A., *Familiar Faces in Unfamiliar Places: Assyrians in the California Heartland 1911–2010* (Bloomington, IN: Xlibris, 2010).

Ishaya, A., *New Lamps for Old: The Assyrians of North Battleford, Canada* (Piscataway, NJ: Gorgias Press, 2010).

Naby, E., 'Almost Family: Assyrians and Armenians in Massachusetts', in M. Mamigonian (ed.), *Armenians of New England: Celebrating a Culture and Preserving a Heritage* (Belmont, MA: Armenian Heritage Press, 2004), 43–52.

Shoumanov, V., *Assyrians in Chicago* (Charleston: Arcadia Publishing, 2001) (with photos).

Western Europe

Al-Rasheed, M., 'Iraqi Assyrian Christians in London: Beyond the Immigrant/Refugee Divide', *Journal of Assyrian Academic Studies* 26:3 (1995), 241–55 <https://www.anthro.ox.ac.uk/sites/default/files/anthro/documents/media/jaso26_3_1995_241_255.pdf>

Al-Rasheed, M., *Iraqi Assyrian Christians in London: The Construction of Ethnicity* (Lewiston, NY and Lampeter: Edwin Mellen Press, 1998).

First Assyrian Church of the East Purchased in Borken, Germany <https://news.assyrianchurch.org/first-assyrian-church-of-the-east-purchased-in-borken-germany/>

Ex-Soviet Union

Russia

Assyrian Orthodox Christians <http://assyrianorthodox.org/>
Assyriens in Russia <https://en.wikipedia.org/wiki/Assyrians_in_Russia> https://www.religion.in.ua/main/1896-assirijskaya-cerkov-vostoka-zarozhdenie.html> (article by Dmitry Kanibolotsky on the Church of the East).
Naby, E., 'Les Assyriens d'Union soviétique', *Cahiers du monde russe et soviétique* 16:3–4 (1974), 445–57.
Sado, S., *Materials for the Bibliographical Dictionary of Assyrians in Russia* (St Petersburg, 2006) (in Russian).
Seleznyov, N. and G. Kessel, 'Bibliography of Syriac and Christian Arabic Studies in Russian, 2016', *Hugoye: Journal of Syriac Studies* 20:1 (2017), 317–31.
Teule, H. and G. Kessel, 'The Mikhail Sado Collection of Syriac Manuscripts in St. Petersburg', in J. P. Monferrer-Sala, H. Teule and S. Torallas Tovar (eds), *Eastern Christians and their Written Heritage, Manuscripts, Scribes and Context*. Eastern Christian Studies 14 (Leuven: Peeters 2012), 43–76 < <https://www.academia.edu/503714/The_Mikhail_Sado_Collection_of_Syriac_Manuscripts_in_St._Petersburg>
The Church of the East in Moscow <http://www.assyrianchurch.ru/> (in Russian).
The journal *Kuhva d'Madinka* (*The Star*) was published in the Soviet Union after the Revolution of 1917 (in Assyrian).
Vartanov, I., *Assyrians in the Siberian Exile (1949–1956)* (Chicago: Center for Assyrian Genocide Studies, 2008).

Transcaucasia

Yacoub, J. and C. Yacoub, *Oubliés de tous. Les Assyro-Chaldéens du Caucase* (Paris: Éd. du Cerf, 2015).

Armenia

Assyrian Stories from the Caucasus <http://www.aina.org/ata/20150417151222.htm>
Assyrians in Armenia <https://en.wikipedia.org/wiki/Assyrians_in_Armenia>
Chantre, E., *Recherches anthropologiques sur les Aïssores ou Chaldéens émigrés en Arménie* (Lyon: Société d'Anthropologie de Lyon, 1891).
Dum-Tragut, J., 'Assyrians in Armenia: An Interdisciplinary Survey', in *From the Oxus River*, 340–53.
Dum-Tragut, J., 'Zwischen den Fronten: Die russisch-orthodoxen Assyrer von Dimitrov (Armenien)', in D. Winkler (ed.), *Syrische Studien* 11 (Berlin: LIT Verlag, 2016), 143–56.

Georgia

Assyrians in Georgia <https://en.wikipedia.org/wiki/Assyrians_in_Georgia> ; <https://www.georgianjournal.ge/society/27675-extraordinary-life-of-a-confessor-who-speaks-the-language-of-jesus-from-nazareth.html>

Before the First World War the Assyriens de Tiflis published the journal *Madinkha Vostok* (*Orient*).

Hellot-Bellier, F., 'Migration des chrétiens d'Azerbaïdjan iranien vers Tiflis au XIXe siècle', in F. Hellot-Bellier and I. Natchkebia (eds), *La Géorgie entre Perse et Europe* (Paris: Harmattan, 2009), 221–35.

Martin-Hisard, B., 'Les treize saints pères. Formation et évolution d'une tradition hagiographique', *Revue des études géorgiennes et caucasiennes* 1 (1985), 141–68; 2 (1986), 75–111.

Osipov, S., *The Assyrians of Tbilissi* (Tbilissi, 2007) (in Russian).

Outtier, B., 'Georgian Christianity', in K. Parry (ed.), *The Blackwell Companion to Eastern Christianity* (Chichester: Wiley-Blackwell, 2007), 137–55, 142–3.

Ukraine

Irina Kondratieva, *Ancient Eastern Churches: Identification, History, Modernity* (Kiev, 2012) (in Ukrainian) <http://religdep.univ.kiev.ua/index.php/vidannya/159-kondrateva-i-v-davni-skhidni-tserkvi-identifikatsiya-istoriya-suchasnist>

Oukktomski, A., 'The Assyrian Church of the East in Ukraine: A Rapide Survey' (in Russian) <https://www.religion.in.ua/main/history/23362-assirijskaya-cerkov-vostoka-v-ukraine-beglyj-obzor.html>

Australia and New Zealand

Holy Apostolic Catholic Assyrian Church of the East, Archdiocese of Australia, New Zealand and Lebanon <www.assyrianchurch.org.au>

India

Holy Apostolic Catholic Assyrian Church of the East in India <https://web.archive.org/web/20050910232213/http://www.cired.org/ace_india.html>

Holy Apostolic Catholic Church of the East-India <http://churchoftheeastindia.org>

Voice of the East (Magazine) Assyrian Church News <https://news.assyrianchurch.org/voice-of-the-east/>

Timeline

Second millennium BC: The formation of Assyria (in the north of Mesopotamia) with the foundation of a powerful kingdom, which later became an empire. In the seventh and eighth centuries BC Assyria controls an area covering the whole or part of several modern countries, namely, Iraq, Syria, Iran, Lebanon, and Turkey. The Assyrian Empire was succeeded by the Neo-Babylonian Empire (626–539 BC) and by the Achaemenian Persian Empire (c. 559–330 BC), which was conquered by Alexander the Great in 330 BC and which was succeeded in turn by the Greek Seleucids (311–64 BC)

247 BC–224 AD: The Parthian Persian Empire

c. 135–130 BC to 216 AD: The small independent kingdom of Osrhoene with its capital at Edessa, where the language spoken was Syriac

27 BC: Beginning of the Roman Empire

First century AD: The foundation of Christianity with the first Christian communities arising in the regions between Antioch, Jerusalem, Alexandria and Rome

c. 52 AD: The Apostle Thomas (d. c. 68) arrives in India and founds churches

224–637: The Sasanian Persian Empire with its capital at Ctesiphon, also the place of the first episcopal see of the Church of the East

309–79: In Persia, the reign of Shapur II and persecutions of the Christians

313: The Edict of Milan promulgated by the emperor Constantine (306–37) officially recognizes the Christian religion in the Roman Empire

325: The Church Council of Nicaea near Constantinople at which the Christian theology of the Trinity is accepted

330–1453: The East Roman or Byzantine Empire with its capital at Constantinople founded by the emperor Constantine in 330

381: The Church Council of Constantinople at which the creed called that of Nicaea-Constantinople is established; the organization of the Church on the principle of five patriarchal Churches, or Pentarchy of patriarchates, namely, Rome, Constantinople, Alexandria, and Antioch in 381, with Jerusalem added in 451

410: The Council of Mar Isaac of the Church of the East

424: The Council of Mar Dadisho, at which the Church of the East proclaims its independence, or autocephaly

431: The ecumenical Church Council of Ephesus

451: The ecumenical Church Council of Chalcedon

484: At the council of Beth Lapat, the Church of the East does not recognize the Council of Ephesus

635–845: The official presence of the Church of the East in China during the Tang dynasty (618–907)

632–661: The reigns of the first caliphs in the Middle East

Seventh century: Arab conquests of Palestine and Syria (in 635 and 636), then of the neighbouring regions, including Egypt (in 639–42), the capital Seleucia-Ctesiphon (in 637) and the Mesopotamian region (in 639–40)

661–750: The Umayyads establish an Arab dynasty of caliphs with their capital at Damascus

750–1258: The Abbasid caliphs establish their capital at Baghdad; the Church of the East transfers its seat to Baghdad under the patriarch Timothy I (780–823)

781: The Xi'an Stele (China) with inscriptions in Syriac and Chinese

830: The creation at Baghdad of the Academy of translation and the sciences, called 'the House of Knowledge', also known by the name of 'the House of Wisdom' (*Beth al-Hekma*)

969–1171: The Fatimid caliphate rules in Egypt and incorporates a large part of North Africa and some of the Middle East

1054: Schism between the Church of Rome and that of Constantinople, together with the patriarchates and Churches linked to Constantinople

1038–1307: The Seljuk empire, a Turco-Persian empire, controlling a vast area extending from the Hindu Kush westwards to Anatolia, and from Central Asia to the Persian Gulf

1095–1291: The military expeditions of the Crusades, organized by the Catholic Church to liberate the Holy Land and Jerusalem, which result in the foundation of the Latin Crusader States

1170–1260: The Ayyubid dynasty in Syria. Its first ruler is Saladin (d. 1193) who occupies Jerusalem in 1187

1190–1222: Yahballaha II, born in China, is enthroned as patriarch of the Church of the East

1206–1368: The Mongol Empire is founded by Genghis Khan (1206–27) and comes to be divided into several 'kingdoms'. Those in the west are called khanates and were assigned to his sons and successors. Among them are the khanate of Jagatai in Central Asia (1242–1347), the ilkhanat ('vassal khanate') of Persia (1256–1335), the Golden Horde (a state to the south of the Volga) governed by a dynasty founded by Jochi, the eldest son of Genghis Khan. In 1226 the Mongols attack the principalities of Rus' (the future Russia), which are obliged to recognize the suzerainty of the Mongols from the thirteenth to the fifteenth century

1240–2: The Mongols of the Golden Horde briefly invade Central Europe (as far as Pest and Neustadt, near Vienna, in 1241 and modern Croatia in 1242)

1250–1517: The Mamluk sultans seize power from their masters the Ayyubid sultans and dominate Egypt (with their capital at Cairo) and Syria

1258: The Mongol army, under the leadership of Hulagu Khan, a grandson of Genghis Khan, captures Baghdad

1260–1368: A new presence of the Church of the East in China under the Mongol Yuan dynasty founded by the great khan Kublai (1260–94) with its capital at Khanbaliq (Beijing)

1281–1317: Under the patriarch Yaballaha III, the Church of the East attains its broadest geographical expansion

1287–8: The diplomatic mission of the monk of the Church of the East, Sauma, to Western Europe

1310: Massacres of Christians at Erbil

1370–1405: Tamerlane conquers a large part of Central and Western Asia, with his capital at Samarkand, and founds the Timurid dynasty, which lasts to 1507

1453: The Ottomans bring the Byzantine Empire to an end with the conquest of Constantinople (called Istanbul since 1930) by Mehmed II in 1453, and dominate the Near East (Syria and Mesopotamia) from 1516; the sultan Selim I conquers Mamluk Egypt in 1517 and his son Suleiman conquers Baghdad and the modern territory of Iraq in 1533

1498: The arrival in India of the Portuguese with Vasco da Gama accompanied by Roman Catholic missionaries

1501–1736: In Persia, the Safavid dynasty succeeds the Timurids

1552–82: The khanate of Kazan is taken by Ivan IV, the first tsar of Russia (1547–84), in 1552 and that of Astrakhan is taken in 1556; in 1582 the beginning of the Russian colonisation of the khanate of Siberia

1553: Jean (Yohannan) Sulaqa, abbot of the monastery of Rabban Hormizd at Alqosh, becomes in Rome the first patriarch of the Church called 'the Chaldean Catholic Church'

1599: The synod of Diamper in India and the creation of the Catholic Syro-Malabar Church

1653: In India, at Coonan, a group of Indian Christians of the Church of the East in Kerala make a vow not to submit to the domination of the Portuguese and the authority of the pope of the Church of Rome

Timeline

1828: After the wars of 1813 and 1826 won by the Russians, the Persians are forced to sign the Treaty of Turkmenchay with the annexation by Russia of Transcaucasia (including Georgia and Armenia)

1830: Yohannan (John) VIII (Hormizd) becomes patriarch of the Catholic Chaldean Church with his seat at Mosul

1843 and 1846: Badr and Nurallah, two Kurdish leaders, attack the Assyrian Christians of the region of Tyari (in the south-east of modern Turkey); massacres of villages in the districts of Lower Tiyari and Tkhuma

1914–15: In the Ottoman Empire the genocide of Armenian, Syriac and Assyro-Chaldean Christians; the Assyrians leave Hakkari with their patriarch

July 1918: The Ottomans take the town of Urmia and massacre thousands of Christians; the survivors decide to leave the town and seek refuge in Iraq, then occupied by the British

1919: At Paris, the international Peace Conference results in the Treaty of Versailles; the victorious allied powers of the First World War (1914–18) negotiate peace treaties between the Allies and the vanquished; several Assyro-Chaldean delegations ask for the creation of a state

1920: At San Remo (Italy), the post-First World War fate of the Arab provinces of the Ottoman Empire is decided; Palestine and Iraq are placed under British mandate and Syria and Lebanon under French mandate; the conference ignores the repatriation of Hakkari's Assyrian Christians and nothing is decided on this matter

24 July 1923: The Treaty of Lausanne replaces the Treaty of Sèvres of 10 August 1920; there is no further discussion of minorities, including the Assyro-Chaldeans

October 1923: The proclamation of the Republic of Turkey, which replaces the Ottoman Empire in Asia Minor, and the country adopts a secular constitution

1925: The League of Nations attaches Hakkari to the new Turkish republic

1932: At the end of the British mandate in Iraq the problem of the re-establishment and repatriation of the Assyrians remains unresolved

7–15 August 1933: In the north of Iraq the massacre of Assyrian Christians at Simele and the surrounding area; about sixty Assyrian villages are pillaged and more or less destroyed

18 August 1933: The Assyrian patriarch Shimun XXI Eshai and his family are expelled from Iraq to Cyprus by the Iraqi government; the patriarchal seat was moved to Chicago (United States) in 1940

1958: In Iraq the monarchy is replaced by a republic

1968: The Church of the East adopts the New Calendar, which creates an internal schism and leads to the formation of the Ancient Church of the East with its seat at Baghdad

1979: The proclamation of the Islamic Republic in Iran

2003: Following the invasion of Iraq by the United States, the president, Saddam Hussein (d. 2006), is overthrown

2005: The regional government of Kurdistan (RGK) in northern Iraq acquires a high degree of autonomy

2014–19: The creation of the caliphate of the Islamic Republic (IS), which makes Mosul its capital in Iraq and Al-Raqqah its capital in Syria until its fall in December 2017 in Iraq and in March 2019 in Syria

2015: The election of Giwargis III as patriarch of the Church of the East; the seat of the patriarchate is moved from Chicago to Erbil

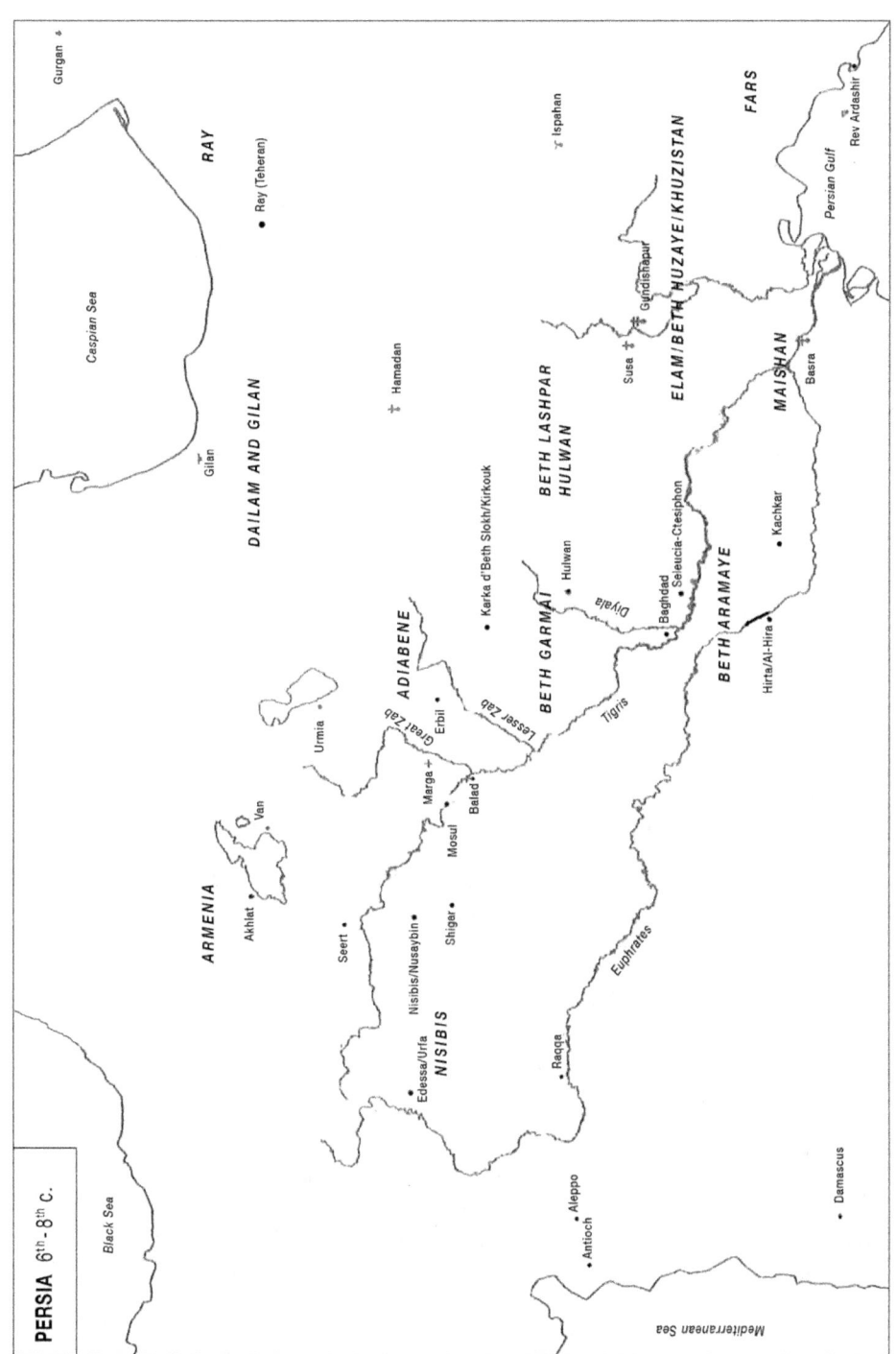

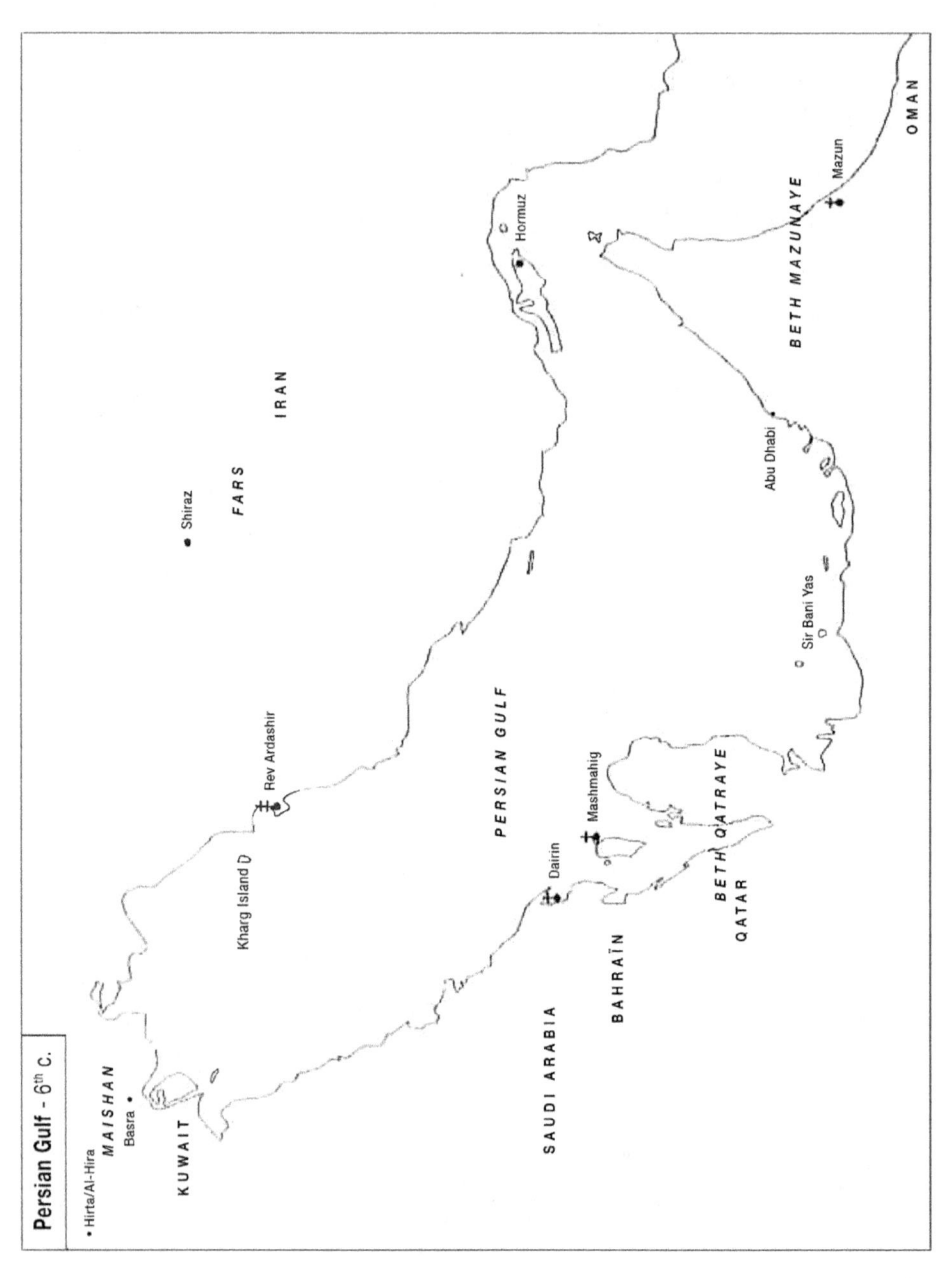

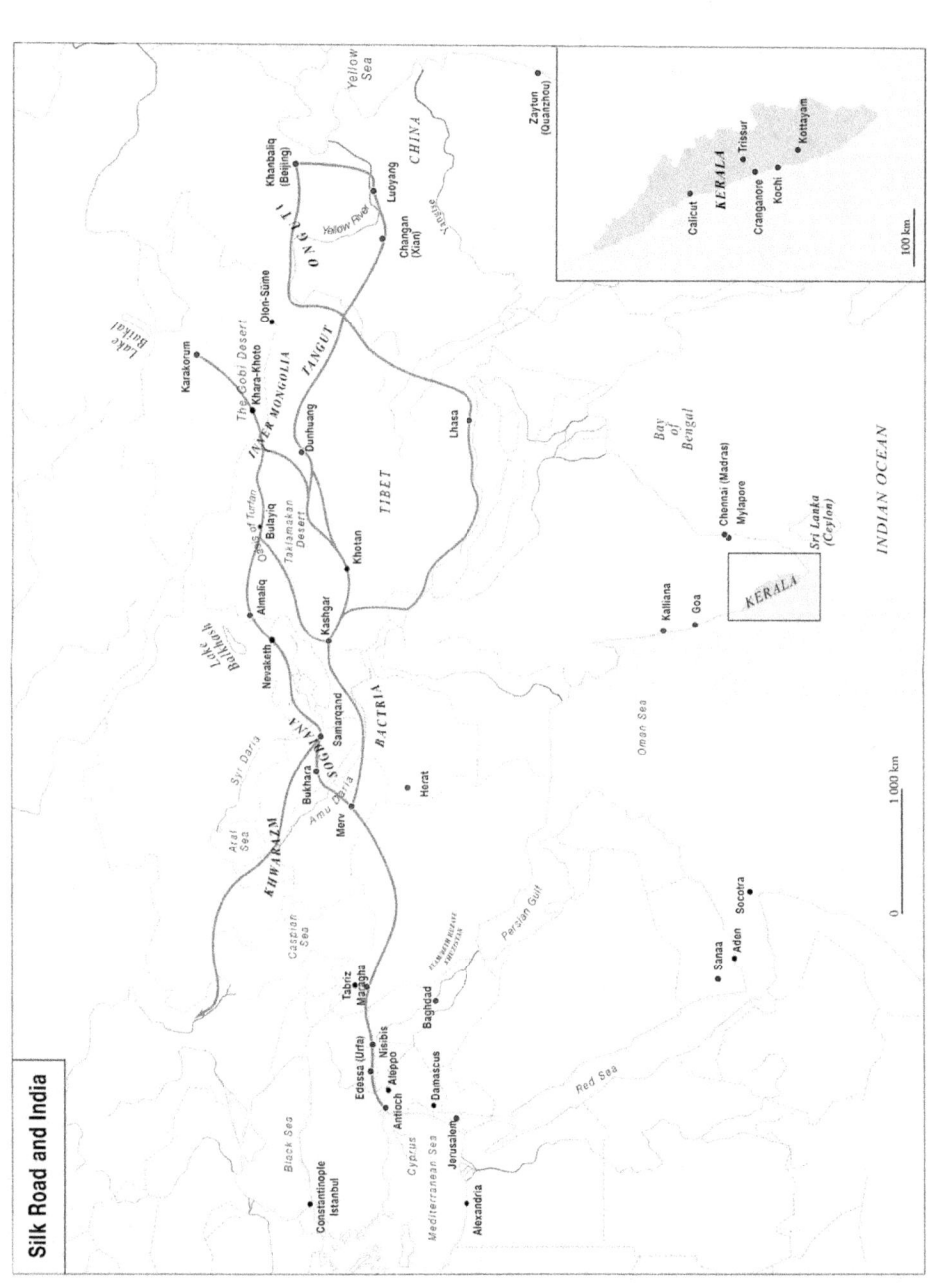

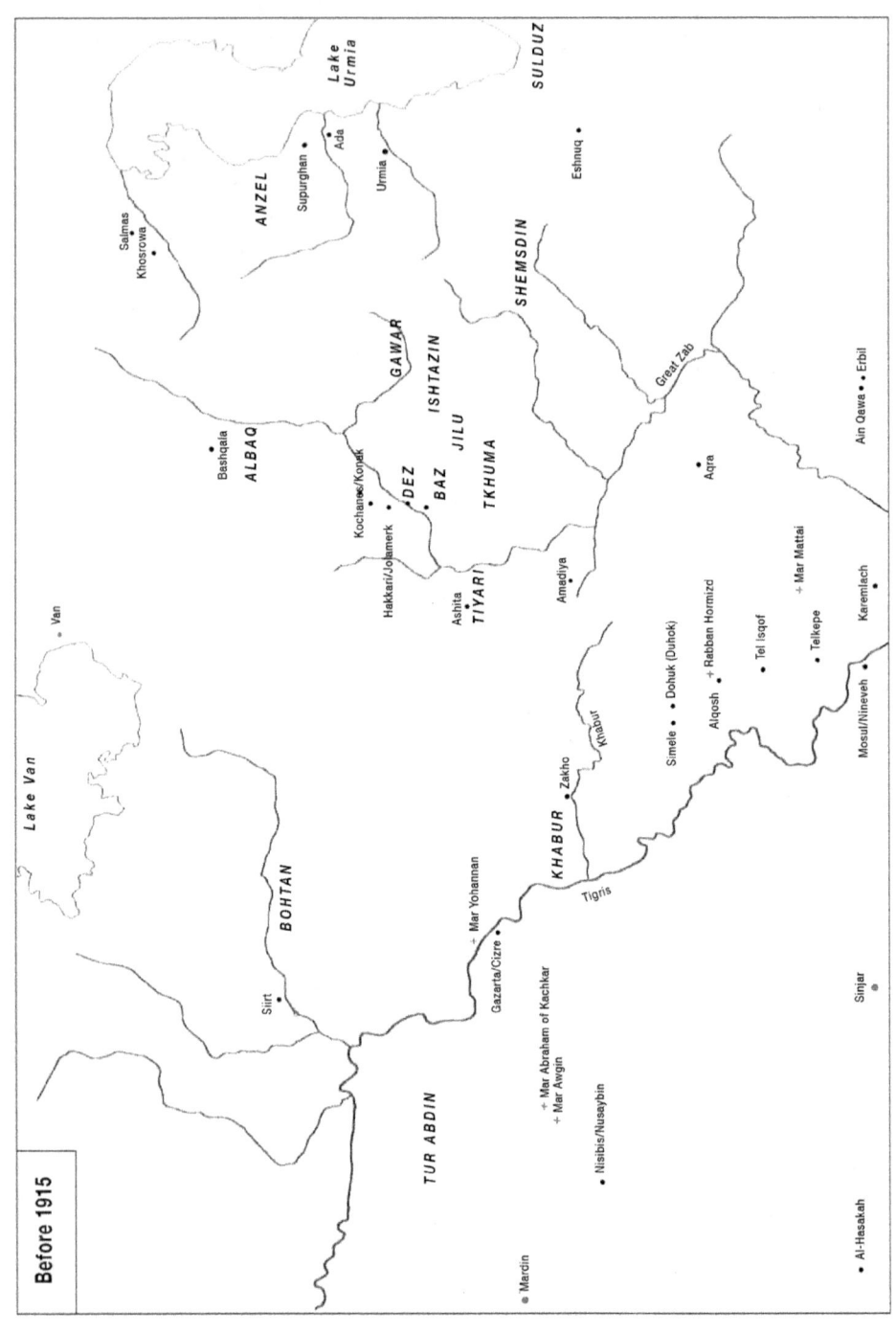

www.ingramcontent.com/pod-product-compliance
Lightning Source LLC
Chambersburg PA
CBHW051737230426
43670CB00012B/2052